Numerical Treatment of the Navier-Stokes Equations

Edited by Wolfgang Hackbusch
and Rolf Rannacher

Notes on Numerical Fluid Mechanics (NNFM)   Volume 30

Series Editors: Ernst Heinrich Hirschel, München
Kozo Fujii, Tokyo
Bram van Leer, Ann Arbor
Keith William Morton, Oxford
Maurizio Pandolfi, Torino
Arthur Rizzi, Stockholm
Bernard Roux, Marseille

(Adresses of the Editors: see last page)

Volume 6   Numerical Methods in Laminar Flame Propagation (N. Peters/J. Warnatz, Eds.)
Volume 7   Proceedings of the Fifth GAMM-Conference on Numerical Methods in Fluid Mechanics (M. Pandolfi/R. Piva, Eds.)
Volume 8   Vectorization of Computer Programs with Applications to Computational Fluid Dynamics (W. Gentzsch)
Volume 9   Analysis of Laminar Flow over a Backward Facing Step (K. Morgan/J. Periaux/F. Thomasset, Eds.)
Volume 10   Efficient Solutions of Elliptic Systems (W. Hackbusch, Ed.)
Volume 11   Advances in Multi-Grid Methods (D. Braess/W. Hackbusch/U. Trottenberg, Eds.)
Volume 12   The Efficient Use of Vector Computers with Emphasis on Computational Fluid Dynamics (W. Schönauer/W. Gentzsch, Eds.)
Volume 13   Proceedings of the Sixth GAMM-Conference on Numerical Methods in Fluid Mechanics (D. Rues/W. Kordulla, Eds.) (out of print)
Volume 14   Finite Approximations in Fluid Mechanics (E. H. Hirschel, Ed.)
Volume 15   Direct and Large Eddy Simulation of Turbulence (U. Schumann/R. Friedrich, Eds.)
Volume 16   Numerical Techniques in Continuum Mechanics (W. Hackbusch/K. Witsch, Eds.)
Volume 17   Research in Numerical Fluid Dynamics (P. Wesseling, Ed.)
Volume 18   Numerical Simulation of Compressible Navier-Stokes Flows (M. O. Bristeau/R. Glowinski/J. Periaux/H. Viviand, Eds.)
Volume 19   Three-Dimensional Turbulent Boundary Layers – Calculations and Experiments (B. van den Berg/D. A. Humphreys/E. Krause/J. P. F. Lindhout)
Volume 20   Proceedings of the Seventh GAMM-Conference on Numerical Methods in Fluid Mechanics (M. Deville, Ed.)
Volume 21   Panel Methods in Fluid Mechanics with Emphasis on Aerodynamics (J. Ballmann/R. Eppler/W. Hackbusch, Eds.)
Volume 22   Numerical Simulation of the Transonic DFVLR-F5 Wing Experiment (W. Kordulla, Ed.)
Volume 23   Robust Multi-Grid Methods (W. Hackbusch, Ed.)
Volume 24   Nonlinear Hyperbolic Equations – Theory, Computation Methods, and Applications (J. Ballmann/R. Jeltsch, Eds.)
Volume 25   Finite Approximations in Fluid Mechanics II (E. H. Hirschel, Ed.)
Volume 26   Numerical Solution of Compressible Euler Flows (A. Dervieux/B. Van Leer/J. Periaux/A. Rizzi, Eds.)
Volume 27   Numerical Simulation of Oscillatory Convection in Low-Pr Fluids (B. Roux, Ed.)
Volume 28   Vortical Solutions of the Conical Euler Equations (K. G. Powell)
Volume 29   Proceedings of the Eighth GAMM-Conference on Numerical Methods in Fluid Mechanics (P. Wesseling, Ed.)
Volume 30   Numerical Treatment of the Navier-Stokes Equations (W. Hackbusch/R. Rannacher, Eds.)

Numerical Treatment of the Navier-Stokes Equations

Proceedings of the Fifth GAMM-Seminar,
Kiel, January 20-22, 1989

Edited by Wolfgang Hackbusch
and Rolf Rannacher

CIP-Titelaufnahme der Deutschen Bibliothek

Numerical treatment of the navier stokes equations:
Kiel, January 20–22, 1989 / ed. by Wolfgang Hackbusch
and Rolf Rannacher. – Braunschweig, Wiesbaden:
Vieweg, 1990
    (Notes on numerical fluid mechanics; Vol. 30)
    (Proceedings of the ... GAMM seminar; 5)
    ISBN 3-528-07630-5

NE: Hackbusch, Wolfgang [Hrsg.]; 1. GT; Gesellschaft
für Angewandte Mathematik und Mechanik:
Proceedings of the ...

Manuscripts should have well over 100 pages. As they will be reproduced photomechanically they should be typed with utmost care on special stationary which will be supplied on request. In print, the size will be reduced linearly to approximately 75 per cent. Figures and diagramms should be lettered accordingly so as to produce letters not smaller than 2 mm in print. The same is valid for handwritten formulae. Manuscripts (in English) or proposals should be sent to the general editor Prof. Dr. E. H. Hirschel, Herzog-Heinrich-Weg 6, D-8011 Zorneding.

Vieweg is a subsidiary company of the Bertelsmann Publishing Group International.

All rights reserved
© Friedr. Vieweg & Sohn Verlagsgesellschaft mbH, Braunschweig 1990

No part of this publication may be reproduced, stored in a retrieval system or transmitted, mechanical, photocopying or otherwise, without prior permission of the copyright holder.

Produced by W. Langelüddecke, Braunschweig
Printed in the Federal Republic of Germany

ISSN 0179-9614

ISBN 3-528-07630-5

Foreword

The GAMM Committee for "Efficient Numerical Methods for Partial Differential Equations" organizes workshops on subjects concerning the algorithmic treatment of partial differential equation problems. The topics are discretization methods like the finite element and the boundary element method for various types of applications in structural and fluid mechanics. Particular attention is devoted to the advanced solution techniques like, e.g., the multigrid method. The series of such workshops was continued in 1989, January 20-22, at the University of Kiel with the 5th Kiel-Seminar on the special topic

"Numerical Treatment of the Navier-Stokes Equations".

The seminar was attended by 66 scientists from 8 countries and 21 lectures were given. The list of topics contained accurate discretization schemes for the compressible as well as the incompressible Navier-Stokes equations, and for certain of its non-viscous approximations, including some self-adaptive mesh generation and extrapolation techniques. Other lectures dealt with the efficient solution of the arising algebraic systems by preconditioned conjugate gradient and multi-grid methods. These proceedings contain 15 of the contributions to the seminar in alphabetical order.

The editors thank the DFG (Deutsche Forschungsgemeinschaft) for its support. They also like to express their gratitude to all persons involved in the organization of the seminar.

November 1989   W. Hackbusch
                                                                                 R. Rannacher

Contents

R. ANSORGE: Extension of an Abstract Theory of Discretization Algorithms to Problems with only Weak and Non-Unique Solutions 1

O. AXELSSON, J. MAUBACH: A Time-Space Finite Element Method for Nonlinear Convection Diffusion Problems 6

P. BASTIAN, G. HORTON: Parallelization of Robust Multi-Grid Methods: ILU Factorization and Frequency Decomposition Method 24

H. BLUM: The Influence of Reentrant Corners in the Numerical Approximation of Viscous Flow Problems 37

M. BORSBOOM: A Finite Discretization with Improved Accuracy for the Compressible Navier-Stokes Equations 47

L. FUCHS: Calculation of Viscous Incompressible Flows in Time-Dependent Domains 62

P. KAPS: Two-Dimensional Wind Flow over Buildings 72

E. KATZER: Laminar Shock/Boundary-Layer Interaction - A Numerical Test Problem 82

B. MÜLLER: Comparison of Upwind and Central Finite-Difference Methods for the Compressible Navier-Stokes Equations 90

E. RIEGER, H. SCHÜTZ, D. WOLTER, F. THIELE: A Comparison of Finite-Difference Approximations for the Stream Function Formulation of the Incompressible Navier-Stokes Equations 100

M.A. SCHMATZ: NSFLEX - An Implicit Relaxation Method for the Navier-Stokes Equations for a Wide Range of Mach Numbers 109

A. SCHÜLLER: A Multigrid Algorithm for the Incompressible Navier-Stokes Equations 124

T.M. SHAH, D.F. MAYERS, J.S. ROLLETT: Analysis and Application of a Line Solver for the Recirculating Flows Using Multigrid Methods 134

R. VERFÜRTH: A Posteriori Error Estimators and Adaptive Mesh-Refinement for a Mixed Finite Element Discretization of the Navier-Stokes Equations 145

G. WITTUM: R-Transforming Smoothers for the Incompressible Navier-Stokes Equations 153

List of Participants 163

Extension of an Abstract Theory of Discretization Algorithms to Problems with only Weak and Non-Unique Solutions

Rainer Ansorge

Institut für Angewandte Mathematik der Universität Hamburg,
Bundesstr. 55, D-2000 Hamburg 13, W-Germany

Assume J to be an index set and X to be a metric space. Let Y be a metric space, too, $\{a(\phi)|\phi \in J\} \subset Y$ a given set and $\{A(\phi)|\phi \in J\}$ a set of operators each of which maps $D = \bigcap_{\phi \in J} D(A(\phi)) \subset X$ into Y; $(D(A(\phi))$: Domain of $A(\phi))$. Consider the problem:

Find $x \in X$ such that
$$A(\phi)x = a(\phi), \quad \forall \phi \in J.^1 \tag{1}$$

Let
$$S := \{x \in X \mid x \text{ solves } (1)\}.$$

The elements of S are called "(weak) solutions". Let (1) be numerically approximated by a sequence of problems
$$\hat{A}_n x_n = \hat{a}_n \quad (n = 1, 2, \ldots) \tag{2}$$

where $\{\hat{A}_n\}$ is a sequence of operators each of which maps X into Z with a certain metric space Z and $\{\hat{a}_n\}$ discretely compact. Let
$$S_n = \{x_n \in X \mid x_n \text{ solves } (2)\} \quad (n = 1, 2, \ldots)$$

and assume

as well as
$$\left. \begin{array}{c} S_n \neq 0 \\ \bigcup_{n \in \mathbb{N}} S_n \text{ bounded} \end{array} \right\}. \tag{3}$$

The solutions x_n $(n = 1, 2, \ldots)$ of (2) are called approximate solutions of (1).

Let $\{A_n(\phi)|\phi \in J\}$ $(n = 1, 2, \ldots)$ be a sequence of sets of operators each of which maps their joint domain $\tilde{D} \subset X$ into Y and let $\{a_n\}$ be a sequence of mappings from J into Y with
$$\lim_{n \to \infty} a_n(\phi) = a(\phi) \quad \text{for every fixed} \quad \phi \in J. \tag{4}$$

Assume
$$x_n \in S_n \Rightarrow A_n(\phi)x_n = a_n(\phi), \forall \phi \in J^2. \tag{5}$$

Then the following theorem holds:

Theorem 1: Let

(i) $[\{A_n(\phi)\}, A(\phi)]$ be discretely closed[3] for every fixed $\phi \in J$.

(ii) $\{\hat{A}_n\}$ be asymptotically regular[3].

---
[1] Normally, (1) is a weak formulation of a problem $Ax = a$.
[2] We call this equation the "weak formulation" of the approximate problem.
[3] For the definition of "discrete closure", "asymptotic regularity", and "discrete compactness (d - compactness)" cf. [1].

Then $\quad\quad\quad\quad\quad S \neq \emptyset \quad\quad$ (existence of a solution of (1))
and $\quad\quad\quad\quad\quad S_n \to S \quad\quad$ in the sense of set convergence.
Proof: $\{\hat{a}_n\}$ discretely compact[3] (i.e. $\hat{A}_n\, x_n,\ x_n \in S_n$, discretely compact) $\Rightarrow \{x_n\}$ d-compact $\Rightarrow \exists\, \{x_n \mid n \in \mathbb{N}' \subset \mathbb{N}\},\ \exists\, x \in X$ with $x_n \to x (n \in \mathbb{N}')$. Herewith, $\mathbb{N}'$ and x are independent of ϕ. Because of $A_n(\phi) x_n = a_n(\phi)$ and $a_n(\phi) \to a(\phi)$, $\forall \phi \in J$, (i) yields $A(\phi) x = a(\phi), \forall \phi \in J$, i.e. $x \in S$. ∎

This theorem generalizes slightly one of our results presented in [1] where theories of Stummel [2] and Grigorieff [3] were generalized.

Remark 1: If one is only interested in particular right hand sides $\hat{a}_n(\phi), a(\phi)$, (ii) can obviously be replaced by the weaker assumption

(ii)* $\quad\quad\quad\quad\quad\quad\quad\quad \{S_n\}$ d-compact.

Often, (1) is completed by an additional inequality constraint of the type

$$B(\phi) x \leq 0, \quad \forall \phi \in \hat{J} \subset J \tag{6}$$

where $\{B(\phi) \mid \phi \in \hat{J}\}$ is a set of nonlinear continuous functionals mapping X into $\mathbb{R}$. Assume an uniqueness theorem to be true:

$$\text{(1) together with (6) has at most one solution.} \tag{7}$$

Because of particular applications in mathematical gas dynamics[4], we call (6) an "entropy condition" and the unique solution $\hat{x}$ of $\{(1), (6)\}$ – if it exists – is called the "entropy solution" of (1).

Let us also complete the approximate problem (2) by inequality constraints

$$B_n(\phi) x_n \leq 0, \quad \forall \phi \in \hat{J} \quad\quad (n = 1, 2, \ldots) \tag{8}$$

with continuous functionals $B_n(\phi)$ and assume that $\{(2), (8)\}$ has at least one solution $\hat{x}_n \in S_n$ for every fixed $n \in \mathbb{N}$. We call (8) a sequence of "discretized entropy conditions", but we do not necessarily expect that $\{(2), (8)\}$ has a unique solution for every fixed n.

Then the following theorem holds:

Theorem 2: Let $\hat{S}_n := \{\hat{x}_n \in S_n \mid \hat{x}_n \text{ fullfills (8), too}\}$ and assume

$$B_n(\phi) \xrightarrow{c} B(\phi) \quad \text{(continuous convergence) for every fixed} \quad \phi \in \hat{J}[5].$$

Then the entropy solution $\hat{x}$ of $\{(1), (6)\}$ exists and

$$\hat{S}_n \to \{\hat{x}\},$$

i.e. every sequence $\{\hat{x}_n\}$ converges to the unique entropy solution of the original problem[6].

Proof: Because of $S_n \to S$ (cf. Theorem 1) there is a subsequence $\{\hat{x}_n \in \hat{S}_n \mid n \in \mathbb{N}' \subset \mathbb{N}\}$ and a solution $x \in S$ with $\hat{x}_n \to x$, $n \in \mathbb{N}'$.

The continuous convergence yields $B_n(\phi) \hat{x}_n \to B(\phi) x$. Because of (8), $B(\phi) x \leq 0$. Hence, x is the entropy solution $\hat{x}$. Because $\hat{x}$ is unique, not only subsequences $\{\hat{x}_n \mid n \in \mathbb{N}'\}$ converge to $\hat{x}$ but every full sequence $\{\hat{x}_n \mid n \in \mathbb{N}\}$ behaves so.

---

[4] cf. the subsequent Example
[5] For the definition of continuous convergence, cf. [1]
[6] even if the $\hat{x}_n$ for fixed n are not unique

It should be mentioned that – due to a theorem of Rinow [4], p. 78 – continuous convergence $B_n(\phi) \xrightarrow{c} B(\phi)$ for fixed $\phi \in \hat{J}$ is equivalent with the pair of properties

(j)   the functionals $B_n(\phi)$ $(n = 1, 2, \ldots; \phi \in \hat{J}$ fixed)
are equicontinuous,   (9)
(jj) $B_n(\phi)x \to B(\phi)x$ pointwise for every fixed $\phi \in \hat{J}$
and for every fixed $x \in \tilde{X} \underset{\text{dense}}{\subset} X$.

Example: Gas flows are mathematically modeled by a system of quasilinear partial differential equations of conservation law form (cf. e.g. [5]). Let us therefore consider a system of p equations of conservation law form in one space variable together with an initial condition, i.e.

$$Ax = \left\{ \begin{array}{c} x_t + \frac{\partial}{\partial \xi} q(x) \\ x(\xi, 0) - x_0(\xi) \end{array} \right\} = 0, \quad (\xi, t) \in \Omega \tag{10}$$

where Ω is the upper half (ξ, t)-plane including the ξ-axis:

$$\Omega = \{(\xi, t) \; \xi \in \mathbb{R}, \; t \geq 0\}.$$

If the Jacobian of q is continuous and if the initial function x_0 is smooth, there is a local existence and uniqueness theorem for problem (10) but applications require the study of global solutions. But because (10) can lead to shocks, one has to deal with weak solutions. The weak formulations of (10) is the following one:

Find $x \in \left[L_1^{\text{loc}}(\Omega) \right]^p =: X$ such that
$$A(\phi)x := - \int_\Omega \int \{\phi_t x + \phi_\xi q(x)\} d\xi dt - \int_\mathbb{R} \phi(\xi, 0) x_0(\xi) d\xi = 0 \tag{11}$$
for all test functions $\phi \in C_0^1(\Omega) =: J$.

Hence, $Y = \mathbb{R}^p$ and $a(\phi) = 0$ for all $\phi \in J$. Smooth solutions of (10) fullfill also (11), and vice versa. But the transition from the smooth problem (10) to the weak formulation (11) leads to a loss of uniqueness.

Lax [6] therefore recommmended the completion of the conservation law system by additional regard of the 2nd Main Theorem of thermodynamics, i.e. the validity of

$$d(-\tilde{S}) \;\; \leq \;\; 0 \tag{12}$$

in closed systems where $\tilde{S}$ is the entropy. Generalization to arbitrary one-dimensional systems of conservation laws leads to the following definition:

$\hat{x}$ is called an entropy solution of (11) if there are functionals
V and F mapping X into $\mathbb{R}$ such that

(k) $V = V(x)$ is convex,
(kk) $V_t + \frac{\partial}{\partial \xi} F(x) = 0$ is automatically fullfilled if
         x is a smooth solution of (11) [7].
(kkk) $V_t(\hat{x}) + \frac{\partial}{\partial \xi} F(\hat{x}) \leq 0$ weakly[8], i.e.

$$B(\phi)\hat{x} \;\; := \;\; - \int_\Omega \int \{\phi_t V(\hat{x}) + \phi_\xi F(\hat{x})\} d\xi dt$$

---

[7]This follows from the fact that smooth solutions are unique and keep the entropy constant. (kk) establishes the connection between V and F.
[8]In the case of gas dynamics, $V = -s$ with $s = \tilde{S}/volume$

$$-\int_{\mathbb{R}} \phi(\xi,0)V(x_0(\xi))d\xi \leq 0, \tag{13}$$

$$\forall \phi \in \{\phi \in J | \phi \geq 0\} =: \hat{J}.$$

V is called "entropy functional".
In order to treat our example (11) numerically, let us introduce equidistant step sizes h and Δt with

$$\frac{h}{\Delta t} =: \lambda = \text{const} > 0, \tag{14}$$

and let $\xi_j = jh$, $t_\nu = \nu \Delta t$. Only for convenience, we restrict ourselves to explicit three point difference schemes

$$x_j^{\nu+1} - x_j^\nu + \lambda \left\{ g_{j+\frac{1}{2}}^\nu - g_{j-\frac{1}{2}}^\nu \right\} = 0 \tag{15}$$

where the "numerical flux"

$$g_{j+\frac{1}{2}}^\nu := g\left(x_j^\nu, x_{j+1}^\nu\right) \tag{16}$$

is assumed to be lipschitz continuous and to be "consistent" with (10), i.e.

$$g(\sigma,\sigma) = q(\sigma), \ \forall \sigma \in \mathbb{R}^p. \tag{17}$$

The discrete initial values are

$$x_j^0 := \frac{1}{h} \int_{(j-\frac{1}{2})h}^{(j+\frac{1}{2})h} x_0(\xi) d\xi \tag{18}$$

where x_0 is assumed to have compact support in $\mathbb{R}$. Assume $h = O(\frac{1}{n})$ $(n \in \mathbb{N})$ and put

$$x_n(\xi,t) := x_j^\nu \text{ for } \begin{cases} (j-\frac{1}{2})h < \xi < (j+\frac{1}{2})h & (j=0,\pm 1, \pm 2, \ldots) \\ \nu \Delta t \leq t < (\nu+1)\Delta t & (\nu = 0,1,2,\ldots). \end{cases} \tag{19}$$

Hence, (15) together with (18) can be written as

$$\hat{A}_n x_n = 0. \tag{20}$$

Moreover, $x_n \in L_1^{\text{loc}}(\Omega) = X$, and the weak formulation of $\{(15), (18)\}$ – which is even equivalent with $\{(15), (18)\}^9$ – reads as

$$\begin{aligned} A_n(\phi)x_n := & \frac{1}{\Delta t} \int\int_\Omega \{x_n(\xi, t+\Delta t) - x_n(\xi,t)\} \phi(\xi,t) d\xi dt \\ & + \frac{1}{h} \int\int_\Omega \{g(x_n(\xi,t), x_n(\xi+h,t)) - g_n(x_n(\xi-h,t), x_n(\xi,t))\} \phi(\xi,t) d\xi dt. \end{aligned} \tag{21}$$

To show the discrete closure of $[\{A_n(\phi)\}, A(\phi)]$ (cf. propertiy (i) of Theorem 1), one has to prove:

$$\left. \begin{array}{ll} x_n & \to \ x \ \text{(with respect to the } L_1\text{-topology)}, \\ A_n(\phi)x_n & \to \ y \ \text{(in } \mathbb{R}) \end{array} \right\} \Rightarrow A(\phi)x = y.$$

But in the context of our example, this is exactly nothing other than the Lax-Wendroff Theorem (cf. [7]).

---

[9] thus, (5) is fullfilled.

Because we are only interested in the particular right hand sides $\hat{a}_n = 0$, $a_n(\phi) = 0$, $a(\phi) = 0$, it is sufficient to show the validity of assumption (ii)* of Remark 1. But if we restrict ourselves to the case $p = 1$, the original problem (11) shows the well-known total variation diminishing property (TVD property) of its weak solutions such that it is reasonable to use approximate problems the solutions of which have this property, too. Such "TVD schemes", i.e. suitable functions g, were constructed by Harten in [8], and it was in this paper where Harten showed that his schemes fullfill (ii)*.

Because of the explicitness of the three point methods under consideration, $S_n \neq 0, \forall n \in \mathbb{N}$. And because the TVD property of the schemes yields the boundedness of $\bigcup_n S_n$, also (3) is fullfilled.

Harten also showed that his numerical solutions converge to the entropy solution of the original problem, but he did this without explicit characterization of discretizations of V and F, i.e. without definition of $B_n(\phi)$ $(n = 1, 2, \ldots)$. But discrete functionals $B_n(\phi)$ which fit Harten's scheme can be described, $B_n(\phi) \stackrel{c}{\to} B(\phi)$ can be shown such that the proof of $x_n \to \hat{x}$ becomes more pertinent and more easy.

References

[1] ANSELONE, P.M. AND ANSORGE, R.: *Discrete closure and asymptotic (quasi-) regularity in discretization algorithms.* IMA J. Num. Anal. 7, 431-448 (1987)

[2] STUMMEL, F.: *Discrete convergence of mappings.* Proceedings of the Conference on Numerical Analysis, Dublin (1972), 285-310. Academic Press 1973

[3] GRIGORIEFF, R.D.: *Über diskrete Approximationen nichtlinearer Gleichungen 1. Art.* Math. Nachr. 69, 253-272 (1975)

[4] RINOW, W.: *Die innere Geometrie der metrischen Räume.* Berlin-Göttingen-Heidelberg: Springer 1961

[5] CHORIN, A.J. AND MARSDEN J.E.: *A mathematical introduction to fluid mechanics (2nd printing).* New York-Berlin-Heidelberg-Tokyo: Springer 1984

[6] LAX, P.D.: *Hyperbolic systems of conservation laws, II.* Comm. Pure Appl. Math. 10, 537-566 (1975)

[7] LAX, P.D. AND WENDROFF, B.: *Systems of conservation laws.* Comm. Pure Appl. Math. 13, 217-237 (1960)

[8] HARTEN, A.: *On a class of high resolution total-variation-stable finite-difference schemes.* SIAM J. Num. Anal. 21, 1-23 (1984)

A Time-Space Finite Element Method for Nonlinear Convection Diffusion Problems

O. Axelsson and J. Maubach[†]

Faculty of Mathematics and Computer Science,
University of Nijmegen
Toernooiveld 5, 6525 ED Nijmegen
The Netherlands

Summary

Time-stepping methods for parabolic problems require a careful choice of the stepsize for stability and accuracy. Even if a stable implicit time-stepping method is used, one might be forced to choose very small time-steps in order to get a sufficient accuracy, if the solution has steep gradients, even if these occur only in a narrow part of the domain. Therefore the solution of the corresponding algebraic systems can be expensive since many time-steps have to be taken. The same considerations are valid for explicit time-stepping methods. In this paper we present a discretization technique in which finite element approximations are used in time and space simultaneously for a relatively large time period called 'time-slab'. This technique may be repeatedly applied to obtain further parts of the solution in subsequent time-intervals. It will be shown that, with the method proposed, the solution can be computed cheaply, even if it has steep gradients, and that stability is automatically guaranteed. For the solution of the non-linear algebraic equations on each time-slab fast iterative methods can be used.

Keywords: Time-stepping, Time-space finite elements, Convex diffusion, Non-linear parabolic differential equations Mesh refinement
Subject Classifications: AMS(MOS): 65F10,65M20,65N30,65N50

[†] research of second author supported by the Netherlands Organization for Scientific Research N.W.O.

1. Introduction

The most frequently used method for the numerical integration of parabolic differential equations is the method of lines, where one first uses a discretization of space derivatives by finite differences or finite elements and then uses some time-stepping method for the solution of the resulting system of ordinary differential equations. Such methods are, at least conceptually, easy to perform. However, they can be expensive if steep gradients occur in the solution, stability must be controlled, and the global error control can be troublesome.

This paper considers a simultaneous discretization of space and time variables for a one-dimensional parabolic equation on a relatively long time interval, called 'time-slab'. The discretization is repeated or adjusted for following 'time-slabs' using continuous finite element approximations. In such a method we utilize the efficiency of finite elements by choosing a finite element mesh in the time-space domain where the finite element mesh has been adjusted to steep gradients of the solution both with respect to the space and the time variables. In this way we solve all the difficulties with the classical approach since stability, discretization error estimates and global error control are automatically satisfied. Such a method has been discussed previously in [3] and [4]. The related boundary value techniques or global time integration for systems of ordinary differential equations have been discussed in several papers, see [12] and the references quoted therein. In [16] a time-space method with discontinuous elements in time has been used, which is based on methods in [17], [18] and [20].

In the present paper a non-linear convection diffusion problem is considered. In section 2 the problem is presented and reformulated as a two-dimensional boundary value problem. In section 3 we formulate the discrete problem and a solution method, and in section 4 we consider the stability and discretization error estimates for the method. Finally in section 5, numerical tests and a discussion of the mesh generation method used, is found, and in section 6 some conclusions are drawn.

2. Parabolic differential equations

Consider the following one-dimensional parabolic non-linear partial differential equation on the time-space interval $\Omega := (0,1) \times (0,\infty)$ (see fig. 1)

$$\begin{cases} -(\varepsilon u_x(x,t))_x + au_x(x,t) + \sigma u_t(x,t) + f(x,t) = 0 & 0 < x < 1, 0 < t < \infty \\ u(x,0) = u_0(x) & 0 < x < 1 \\ u(0,t) = l(t) & 0 \leq t < \infty \\ u(1,t) = r(t) & 0 \leq t < \infty \end{cases} \quad (2.1)$$

where the diffusion ε and the flow velocity functions a, σ satisfy $\varepsilon \equiv \varepsilon(u_x^2(x,t))$ resp. $a = a(x,t)$ and $\sigma = \sigma(x,t)$, f is a source function and u_0 some $L^2((0,1))$ integrable function. In addition we assume that $\sigma \geq \sigma_0 > 0$, $a, \sigma \in \tilde{C}^1(\Omega)$, the vector space of continuously differentiable functions, which can be, including their partial derivatives, extended continuously to the boundary of the domain Ω. Further, let $\nabla \cdot (a, \sigma) \equiv a_x + \sigma_t \leq 0$ and define $\varepsilon' \equiv \frac{\partial}{\partial \zeta}\varepsilon(\zeta)$. We assume that $\varepsilon' \geq 0$ The parabolic problem above occurs in many applications of which one was considered in [9].

In the classical way of solving (2.1), one first discretizes the space-variable x, e.g., with the use of a finite element method. Then the calculation of the solution of the system of ordinary differential equations obtained is done with the use of a time-stepping method. One of the disadvantages of this approach is that, in order to get a good approximation of the solution $\hat{u}(x,t)$ for large values of $t > 0$, many small time-steps must be used if the solution has steep gradients, even if these occur only in a small part of the space interval. Furthermore, for explicit time-stepping methods, the stepsize must be chosen to satisfy the Euler method type stability criterion. (However, as shown in [14] and [19] there exist methods with extended stability regions which can partly alleviate this difficulty). Also the local discretization errors made with the use of a time-stepping method have to be monitored closely to control the global errors made in time. Here a method is considered where a finite element mesh is chosen for the time-space domain where these disadvantages do not occur.

In order to compute the solution of (2.1) the computational domain Ω is split up into a number of equidistant 'time-slabs' $\Omega_{t_0} = (0,1) \times (t_0, t_0 + T)$ for $t_0 = 0, T, 2T, \ldots$ (see fig. 2), with lower and upper boundaries denoted by Γ_1 resp. Γ_3, and left and right boundaries Γ_4 resp. Γ_2. The number of such time-slabs is finite, independent of the choice of the mesh parameter, associated with the finite elements. For the first time-slab Ω_0 an initial value u_0 on Γ_1 has to be given, but for each following time-slab Ω_{t_0+T} the solution at Γ_3 of Ω_{t_0} will be taken to provide a Dirichlet boundary condition at Γ_1.

With this approach problem (2.1) can be rewritten into:

$$\begin{cases} -\nabla \cdot (A\underline{\nabla} u(x,t)) + \underline{b}^t \underline{\nabla} u(x,t) + f(x,t) & = 0 & \text{on } \Omega_{t_0} \\ u(x,0) & = u_0(x) & \text{on } \Gamma_1 \\ u(0,t) & = l(t) & \text{on } \Gamma_4 \\ u(1,t) & = r(t) & \text{on } \Gamma_2 \end{cases}, \qquad (2.2)$$

on each time-slab with

- time-slab Ω_{t_0} given by $(0,1) \times (t_0, t_0 + T)$, some time period $0 < T < \infty$ and $t_0 \in \{0, T, 2T, \ldots\}$
- tensor $A \equiv A(u) = \begin{bmatrix} \varepsilon(u_x^2) & 0 \\ 0 & 0 \end{bmatrix}$ and flow field $\underline{b} \equiv \begin{bmatrix} b_1(x,t) \\ b_2(x,t) \end{bmatrix} = \begin{bmatrix} a \\ \sigma \end{bmatrix}$, where we assume that $b_2 \geq b_0 > 0$ in order to preserve the parabolic nature of the equation
- square integrable functions l and r, prescribing the Dirichlet boundary conditions on the left respectively right boundary
- the divergence operator $\nabla\cdot$ and gradient operator $\underline{\nabla}$ defined on the two-dimensional (x,t) space and
- square integrable source function f at Ω and initial value function u_0 at Γ_1.

Note that there is no need to impose any boundary condition at the boundary Γ_3, because for all possible trial functions u and all test functions v the corresponding boundary integral

$$\oint_{\Gamma_3} v(A\underline{\nabla}u)^t \vec{n}\, ds = \oint_{\Gamma_3} v \begin{bmatrix} \varepsilon(u_x^2)u_x \\ 0 \end{bmatrix}^t \begin{bmatrix} 0 \\ 1 \end{bmatrix} ds = 0,$$

where $\vec{n}$ is the outward normal of the boundary $\partial\Omega$. At this boundary the solution $\hat{u}$ of (2.2) and $(\underline{\nabla}\hat{u})^t \vec{n}$ are initially unknown.

An advantage of the formulation (2.2) is that it permits, using small sized elements, for an accurate time-space finite element discretization inside layers, which can arise for $a > 0$ along the boundary Γ_2 (see figs. 6.1 and 6.2), for $a < 0$ along the boundary Γ_4 and in the interior along a shockwave, typically eminating from the south-west corner, if $u_0(0) \neq l(0)$ and $a > 0$ (see fig. 6.3). In other parts of the time-space domain one can use much larger elements thus reducing the number of degrees of freedom considerably compared to a classical time-stepping method.

As we shall see, the computation of the finite element solution on each time-slab can be done efficiently. The solution $\hat{u}$ of (2.2) will be calculated by a non-linear iterative method, which implies that an initial solution u^0 must be provided. If there is any a priori knowledge about the solution then this information can be used to construct a proper initial mesh for each time-slab.

3. Solution method

Consider the variational formulation of the non-linear two-dimensional problem (2.2) for a certain time-slab $\Omega := \Omega_{t_0}$. Let $H^1(\Omega)$ be the Sobolev space of order 1 on Ω and define the boundary function γ at $\Gamma_D := \Gamma_{1,2,4}$ by $\gamma := (u_0, r, l)$, i.e., $\gamma \equiv u_0$ at Γ_1, $\gamma \equiv r(t)$ at Γ_2 and $\gamma \equiv l(t)$ at Γ_4. To simplify the analysis we shall assume that there exists an extension of γ to Ω in $H^1(\Omega)$, which excludes the occurance of interior layers due to discontinuous boundary data. Define the test and trial spaces by $H_0^1(\Omega) := \{v \in H^1(\Omega) : v \equiv 0 \text{ at } \Gamma_D\}$ resp. $H_\gamma^1(\Omega) := \{u \in H^1(\Omega) : u \equiv \gamma \text{ at } \Gamma_D\}$, both in the sense of traces. Now the variational formulation becomes

$$< F(u), v > = 0 \quad \forall_{v \in H_0^1(\Omega)}, u \in H_\gamma^1(\Omega) \tag{3.1}$$

where for fixed $\alpha \geq 0$ the gradient F is given by

$$< F(u), v > = \int_\Omega (-\nabla \cdot (A\nabla u) + \underline{b}^t \nabla u + f) v e^{-\alpha(t-t_0)} \, d\Omega \quad \forall_{u, v \in H^1(\Omega)}. \tag{3.2}$$

Here $t \to e^{-\alpha(t-t_0)}$ is a weight function, which can be useful to get better estimates of the discretization errors, as we shall see in the next section. It suffices to consider the first time-slab where $t_0 = 0$, thus reducing the weight function to $t \to e^{-\alpha t}$.

In this case we have for all $u, v \in H^1(\Omega)$

$$\begin{aligned}
< F(u), v > &= \int_\Omega (A\nabla u)^t \nabla(v e^{-\alpha t}) + (\underline{b}^t \nabla u + f) v e^{-\alpha t} \, d\Omega - \oint_{\partial\Omega} v e^{-\alpha t} (A\nabla u)^t \vec{n} \, ds \\
&= \int_\Omega \varepsilon u_x \frac{\partial}{\partial x}(v e^{-\alpha t}) + (\underline{b}^t \nabla u + f) v e^{-\alpha t} \, d\Omega - \oint_{\Gamma_{2,4}} v e^{-\alpha t} (A\nabla u)^t \vec{n} \, ds \\
&= \int_\Omega \varepsilon u_x v_x e^{-\alpha t} + (\underline{b}^t \nabla u + f) v e^{-\alpha t} \, d\Omega - \oint_{\Gamma_{2,4}} v e^{-\alpha t} (A\nabla u)^t \vec{n} \, ds \\
&= \int_\Omega \left[(A\nabla u)^t \nabla v + (\underline{b}^t \nabla u + f) v \right] e^{-\alpha t} \, d\Omega - \oint_{\Gamma_{2,4}} v e^{-\alpha t} (A\nabla u)^t \vec{n} \, ds
\end{aligned}$$

because $(A\nabla u)^t \vec{n} \equiv 0$ at $\Gamma_{1,3}$ and $\frac{\partial}{\partial x}(v e^{-\alpha t}) = v_x e^{-\alpha t}$.

Linearization of this weak formulation with the use of a damped Newton method now leads to a sequence of linear systems and solutions $u^{k+1} \in H_\gamma^1(\Omega)$

$$< H(u^k)(u^{k+1} - u^k), v > = -\tau_k < F(u^k), v > \quad \forall_{v \in H_0^1(\Omega)}. \tag{3.3}$$

Here the Hessian or Jacobian matrix H is defined by

$$< H(u)w, v > := \lim_{\tau \to 0} \frac{1}{\tau} < F(u + \tau w) - F(u), v > \quad \forall_{u, v, w \in H^1(\Omega)}$$

whence for all $u, v, w \in H^1(\Omega)$

$$< H(u)w, v > = \int_\Omega \left[(B\nabla w)^t \nabla v + \underline{b}^t \nabla w v \right] e^{-\alpha t} \, d\Omega - \oint_{\Gamma_{2,4}} v e^{-\alpha t} (B\nabla w)^t \vec{n} \, ds \tag{3.4}$$

where the tensor B (see [8] for the derivation of B for multi-dimensional problems) is defined by

$$B \equiv B(u) = \begin{bmatrix} \varepsilon(u_x^2) + 2u_x^2 \varepsilon'(u_x^2) & 0 \\ 0 & 0 \end{bmatrix}.$$

Further τ_k is a positive scalar which is called damping parameter (for values less than 1). This scalar can be monitored from step to step in order to achieve convergence, see for instance [1] and [11].

The fact that $u^{k+1} - u^k \in H_0^1(\Omega)$, a linear vector space on which the Hessian will be positive definite (see below), implies that the damped Newton algorithm (3.3) will converge for properly chosen damping parameters τ_k (see e.g. [1]).

Note that, due to the convective term $\underline{b}^t \underline{\nabla} w v$ in the integrand of (3.4), the Hessian matrix $H(u)$ is not symmetric, but because of the special structure of the tensor ε and the nonsymmetric term, the technique described in [8] for symmetric problems, can be modified easily in order to assemble the gradient and Hessian cheaply.

Define the Hilbert space $\hat{H}^1(\Omega) \supset H^1(\Omega)$, the closure of $\tilde{C}^1(\Omega)$ under the weighted norm

$$|||v||| := \left(\int_\Omega (v^2 + v_x^2) e^{-\alpha t} d\Omega \right)^{\frac{1}{2}} \quad \alpha \in \mathbb{R},$$

which is related to a corresponding inner product. Let the norms $\|.\|_s$ and $|.|_s$ denote the Sobolev norm resp. seminorm of order s on $H^1(\Omega)$ whence $|||.||| \leq \|.\|_1$ for all $\alpha \geq 0$. $|||.|||$ can be seen as a weighted Sobolev 1 measure in space combined with a weighted L^2 measure in time on $\hat{H}^1(\Omega)$. With the use of the set of norms introduced and under some assumptions to be derived on the tensor ε and flow field $\underline{b}$, $H(u)$ will be seen to be uniformly positive definite on $H_0^1(\Omega)$, i.e.,

$$<H(u)v,v> \geq c |||v||| > 0 \quad \forall_{v \subset H_0^1(\Omega)},$$

for some positive scalar c.

In order to see this, first note that

$$<H(u)w,v> = \int_\Omega \left[(B\underline{\nabla}w)^t \underline{\nabla}v + \underline{b}^t \underline{\nabla}w v \right] e^{-\alpha t} d\Omega \quad \forall_{v \in H_0^1(\Omega)} \forall_{u,w \in H^1(\Omega)}.$$

An analysis of the separate terms in this expression shows that

$$0 < \lambda_{\min} \int_\Omega v_x^2 e^{-\alpha t} d\Omega \leq \int_\Omega (B(u)\underline{\nabla}v)^t \underline{\nabla}v e^{-\alpha t} d\Omega \quad \forall_{u,v \in H^1(\Omega)}, \tag{3.5}$$

where $\lambda_{\min} := \inf\{\varepsilon(\zeta) + 2\zeta \varepsilon'(\zeta) : \zeta = u_x^2(x,t), (x,t) \in \Omega\}$, and that

$$\int_\Omega \underline{b}^t \underline{\nabla} w v e^{-\alpha t} d\Omega = \oint_{\partial \Omega} v w e^{-\alpha t} \underline{b}^t \underline{n} \, ds - \int_\Omega w \nabla \cdot (\underline{b} v e^{-\alpha t}) d\Omega$$

$$= \oint_{\Gamma_3} v w e^{-\alpha t} \underline{b}^t \underline{n} \, ds - \int_\Omega w (\nabla \cdot \underline{b} v e^{-\alpha t} + \underline{b}^t \underline{\nabla}(v e^{-\alpha t})) d\Omega$$

$$= e^{-\alpha T} \oint_{\Gamma_3} v w \underline{b}^t \underline{n} \, ds - \int_\Omega w (\nabla \cdot \underline{b} v e^{-\alpha t} + \underline{b}^t \underline{\nabla}(e^{-\alpha t}) v + \underline{b}^t \underline{\nabla}(v) e^{-\alpha t}) d\Omega$$

$$= e^{-\alpha T} \oint_{\Gamma_3} v w b_2 \, ds - \int_\Omega \underline{b}^t \underline{\nabla} v w e^{-\alpha t} d\Omega + \int_\Omega v w (\alpha b_2 - \nabla \cdot \underline{b}) e^{-\alpha t} d\Omega$$

$$\forall_{w \in H^1(\Omega)}, \forall_{v \in H_0^1(\Omega)} \tag{3.6}$$

because $v \equiv 0$ at $\Gamma_{1,2,4}$, $\underline{\nabla} e^{-\alpha t} = \begin{bmatrix} 0 \\ -\alpha e^{-\alpha t} \end{bmatrix}$ and $\underline{b}^t \underline{n} = b_2$ at Γ_3. This latter relationship leads to

$$\int_\Omega (\underline{b}^t \underline{\nabla} v) v e^{-\alpha t} d\Omega = \frac{1}{2} e^{-\alpha T} \oint_{\Gamma_3} v^2 b_2 \, ds + \frac{1}{2} \int_\Omega v^2 (\alpha b_2 - \nabla \cdot \underline{b}) e^{-\alpha t} d\Omega$$

$$\geq \frac{1}{2} e^{-\alpha T} \oint_{\Gamma_3} v^2 b_2 \, ds + b_{\min} \int_\Omega v^2 e^{-\alpha t} d\Omega \quad \forall_{v \in H_0^1(\Omega)}$$

where $b_{\min} := \inf\{\frac{1}{2}(\alpha b_2(x,t) - \nabla \cdot \underline{b}(x,t)) : (x,t) \in \Omega\}$. Now (3.5) and the above show that the Hessian satisfies

$$<H(u)v,v> \geq \lambda_{\min} \int_\Omega v_x^2 e^{-\alpha t} d\Omega + b_{\min} \int_\Omega v^2 e^{-\alpha t} d\Omega + \frac{1}{2} e^{-\alpha T} \oint_{\Gamma_3} v^2 b_2 \, ds$$

$$\geq \min\{\lambda_{\min}, b_{\min}\} \int_\Omega (v^2 + v_x^2) e^{-\alpha t} d\Omega \tag{3.7}$$

$$=: c \cdot |||v||| \quad \forall_{v \in H_0^1(\Omega)} \forall_{u \in H^1(\Omega)},$$

i.e., is uniformly positive definite if λ_{min} and b_{min} both are positive. For positive b_{min} this estimate turns out to be uniform in ε.

In the situation where $b_{min} = 0$ note that for piecewise continuous functions v on Ω the restriction to a certain time $t \in (0, T)$, denoted by $v^t: x \mapsto v(x,t)$ will also be piecewise continuous on $(0, 1)$, in particular $v^t \in H^1(0, 1)$. Due to a Friedrichs inequality (see e.g. [21], page 20) there exists a positive scalar $\beta > 0$, not depending on v^t, such that,

$$v^2(1,t) + v^2(0,t) + \int_0^1 (v(x,t))_x^2 \, dx = (v^t(1))^2 + (v^t(0))^2 + \int_0^1 (v^t(x))_x^2 \, dx$$
$$= \oint_{\partial([0,1])} (v^t(s))^2 \, ds + \int_0^1 (v^t(x))_x^2 \, dx$$
$$\geq \beta \cdot \int_0^1 (v^t(x))^2 + (v^t(x))_x^2 \, dx$$
$$= \beta \cdot \int_0^1 (v(x,t))^2 + (v(x,t))_x^2 \, dx \, .$$

Because v is piecewise continuous on Ω and due to the fact that the space-domain does not vary within time, we can integrate the expression above with respect to the time, with the use of a weight $e^{-\alpha t}$, leading to

$$\oint_{\Gamma_2} v^2 e^{-\alpha t} \, ds + \oint_{\Gamma_4} v^2 e^{-\alpha t} \, ds + \int_\Omega v_x^2 e^{-\alpha t} \, d\Omega \geq \beta \int_\Omega (v^2 + v_x^2) e^{-\alpha t} \, d\Omega$$

for all piecewise continuous functions v on Ω. This implies that

$$\left(\int_\Omega v_x^2 e^{-\alpha t} \, d\Omega \right)^{\frac{1}{2}} \text{ and } \left(\int_\Omega (v^2 + v_x^2) e^{-\alpha t} \, d\Omega \right)^{\frac{1}{2}}$$

are equivalent norms on the subspace of piecewise continuous functions of $v \in H^1(\Omega)$ with $v \equiv 0$ at $\Gamma_{2,4}$, whence for $b_{min} = 0$ and such functions v

$$< H(u)v, v > \geq \lambda_{min} \int_\Omega v_x^2 e^{-\alpha t} \, d\Omega + \frac{1}{2} e^{-\alpha T} \oint_{\Gamma_3} v^2 b_2 \, ds$$
$$\geq \beta \lambda_{min} \int_\Omega (v^2 + v_x^2) e^{-\alpha t} \, d\Omega =: c |||v||| \quad \forall_{\alpha \geq 0} \, . \tag{3.8}$$

Because $b_{min} = 0$ this estimate is not bounded uniformly in ε (see the definition of λ_{min}) but contrary to the previous estimate (3.7) it is also valid for $\alpha = b_{min} = 0$. Hence (3.7) will be used mainly for singularly perturbed problems while (3.8) will be used for regular problems.

Each linear system (3.3) is discretized with the use of two-dimensional triangular finite elements (FE) (maximum height h) with linear basisfunctions (see e.g. [23], [13] or [6]). Additional upwind, i.e., streamline-upwind diffusion basisfunctions (SUPG) (see [15] or [5]), is optional. For this latter method the linear basisfunctions v are replaced by $v + \delta \nabla v$ for some scalar $\delta > 0$. As has been shown in [7], for instance, the upwind technique can be very helpful to get a more strongly positive definite system for convection dominated problems and hence increase the rate of convergence of certain generalized preconditioned conjugate gradient iterative methods.

The use of a finite dimensional subspace of $H^1(\Omega)$ in (3.3) to approximate $\hat{u}$ leads to a sequence of corresponding finite dimensional linear systems of the form $H_h(\underline{u}_h^k)(\underline{u}_h^{k+1} - \underline{u}_h^k) = -\tau_k F_h(\underline{u}_h^k)$, defined as usual in finite element methods, with $\lim_{k \to \infty} \underline{u}_h^k := \hat{u}_h$, the discrete solution (see e.g. [1]). These linear non-symmetric finite dimensional systems of equations are solved by iteration with the use of preconditioned linear equation solvers (in the tests GCGLS [2] or CGS [22] are used to this end).

Thus, to approximate the solution of (3.1) we have to provide two stopping criteria, one for the outer non-linear iterations, where $\hat{u}_h$ is approximated,

$$\|F_h(\underline{u}_h^k)\|_e < \varepsilon_{\text{non-linear}},$$

and one for the inner linear iterations where $\underline{u}_h^{k+1}$ is approximated

$$\|H_h(\underline{u}_h^k)(\underline{u}_h^{k+1} - \underline{u}_h^k) + \tau_k F_h(\underline{u}_h^k)\|_e < \varepsilon_{\text{linear}}$$

(here $\|\cdot\|_e$ stands for the Euclidian norm). If the problem is linear we take $\varepsilon_{\text{linear}}$ to be a small positive constant (there will only be one 'non-linear' iteration), and if the problem is non-linear we take $\varepsilon_{\text{linear}} := \rho \|F_h(\underline{u}_h^k)\|_e$ where $\rho := \min(\frac{1}{10}, \|F_h(\underline{u}_h^k)\|_e)$ in order to assure quadratic convergence (see e.g. [1]). For $\varepsilon_{\text{non-linear}}$ a small positive constant is taken. Similarly the damping parameter τ_k is chosen as close to one as possible.

4. Discretization error estimate and stability criteria

To study the discretization error we introduce V_h, a finite dimensional subspace of $H^1(\Omega)$ of test functions, $H_h^1(\Omega) := H^1(\Omega) \cap V_h$, $H_{0,h}^1(\Omega) := H_0^1(\Omega) \cap V_h$, $H_{\gamma,h}^1(\Omega) := H_\gamma^1(\Omega) \cap V_h$. The function γ, which describes the Dirichlet boundary conditions, is assumed to be of such a type that $H_{\gamma,h}^1(\Omega) \neq \emptyset$, e.g., if V_h is a space of piecewise linear functions, then γ has to be piecewise linear too. Also consider the following definitions.

DEFINITIONS
- $\hat{u} \in H_\gamma^1(\Omega)$, a solution of (3.1), i.e., $< F(\hat{u}), v > = 0 \quad \forall_{v \in H_0^1(\Omega)}$,
- $\hat{u}_h \in H_{\gamma,h}^1(\Omega)$, a discrete solution satisfying $< F(\hat{u}_h), v_h > = 0 \quad \forall_{v_h \in H_{0,h}^1(\Omega)}$,
- u_{I_h}, the interpolation of $\hat{u}$ on $H_h^1(\Omega)$,
- $\theta := \hat{u} - \hat{u}_h \in H^1(\Omega)$, the discretization error,
- $\eta := \hat{u} - u_{I_h} \in H^1(\Omega)$, the interpolation error and
- $\zeta := \hat{u}_h - u_{I_h} \in H_{0,h}^1(\Omega)$, the interpolation minus the discretization error.

In order to estimate the discretization error note that $\zeta \equiv 0$ at Γ_D, and assume that ε, $\underline{b}$ with $b_2 > 0$ and $\alpha \geq 0$ are such that for all $u \in H^1(\Omega)$ the following four conditions are satisfied

$$\begin{cases} \lambda_{\min} = \inf\{\varepsilon(\zeta) + 2\zeta\varepsilon'(\zeta) : \zeta = u_x^2(x,t), (x,t) \in \Omega\} > 0 \\ \lambda_{\max} = \sup\{\varepsilon(\zeta) + 2\zeta\varepsilon'(\zeta) : \zeta = u_x^2(x,t), (x,t) \in \Omega\} < \infty \\ b_{\min} = \inf\{\frac{1}{2}(\alpha b_2(x,t) - \nabla \cdot \underline{b}(x,t)) : (x,t) \in \Omega\} > 0 \\ b_{\max} = \sup\{\max\{|b_1(x,t)|, b_2(x,t)\} : (x,t) \in \Omega\} < \infty \end{cases} \quad (4.1)$$

Then for $b_{\min} > 0$ with the use of (3.7)

$$< F(\hat{u}_h) - F(u_{I_h}), \zeta > = < \int_0^1 H(u_{I_h} + \tau\zeta)\zeta \, d\tau, \zeta >$$

$$= \int_0^1 \int_\Omega \left[(B(u_{I_h} + \tau\zeta)\underline{\nabla}\zeta)^t \underline{\nabla}\zeta + \underline{b}^t \underline{\nabla}\zeta \zeta \right] e^{-\alpha t} \, d\Omega \, d\tau$$

$$\geq \lambda_{\min} \int_\Omega \zeta_x^2 e^{-\alpha t} \, d\Omega + b_{\min} \int_\Omega \zeta^2 e^{-\alpha t} \, d\Omega + \frac{1}{2} e^{-\alpha T} \oint_{\Gamma_3} \zeta^2 b_2 \, ds$$

$$\geq \min\{\lambda_{\min}, b_{\min}\} \|\|\zeta\|\|^2 =: c_1 \|\|\zeta\|\|^2$$

or for $b_{min} = 0$ with the use of (3.8)

$$< F(\hat{u}_h) - F(u_{I_h}), \zeta > \geq \lambda_{min} \int_\Omega \zeta_x^2 e^{-\alpha t} d\Omega + \frac{1}{2} e^{-\alpha T} \oint_{\Gamma_3} \zeta^2 b_2 \, ds$$

$$\geq \beta \lambda_{min} \int_\Omega (\zeta^2 + \zeta_x^2) e^{-\alpha t} d\Omega =: c_1 |||\zeta|||^2 \, .$$

Further
$$< F(\hat{u}_h) - F(u_{I_h}), v_h > = < F(\hat{u}) - F(u_{I_h}), v_h > \quad \forall_{v_h \in H_h^1(\Omega)}$$

and for all $\alpha \geq 0$

$$< F(\hat{u}) - F(u_{I_h}), \zeta > = < \int_0^1 H(u_{I_h} + \tau\eta)\eta \, d\tau, \zeta >$$

$$= \int_0^1 \int_\Omega \left[(B(u_{I_h} + \tau\eta)\underline{\nabla}\eta)^t \underline{\nabla}\zeta + \underline{b}^t \underline{\nabla}\eta\zeta \right] e^{-\alpha t} d\Omega \, d\tau$$

$$\leq \int_\Omega \lambda_{max} \left|\eta_x e^{-\frac{1}{2}\alpha t}\right| \left|\zeta_x e^{-\frac{1}{2}\alpha t}\right| d\Omega + \int_\Omega \left|\underline{b}^t \underline{\nabla}\eta e^{-\frac{1}{2}\alpha t}\right| \left|\zeta e^{-\frac{1}{2}\alpha t}\right| d\Omega$$

$$\leq \lambda_{max} (\int_\Omega \eta_x^2 e^{-\alpha t} d\Omega)^{\frac{1}{2}} (\int_\Omega \zeta_x^2 e^{-\alpha t} d\Omega)^{\frac{1}{2}} + b_{max} (\int_\Omega (\eta_x^2 + \eta_t^2) e^{-\alpha t} d\Omega)^{\frac{1}{2}} (\int_\Omega \zeta^2 e^{-\alpha t} d\Omega)^{\frac{1}{2}}$$

$$\leq \max\{\lambda_{max}, b_{max}\} (\int_\Omega (\eta_x^2 + \eta_t^2) e^{-\alpha t} d\Omega)^{\frac{1}{2}} \cdot ((\int_\Omega \zeta_x^2 e^{-\alpha t} d\Omega)^{\frac{1}{2}} + (\int_\Omega \zeta^2 e^{-\alpha t} d\Omega)^{\frac{1}{2}})$$

$$\leq \sqrt{2} \max\{\lambda_{max}, b_{max}\} (\int_\Omega (\eta_x^2 + \eta_t^2) e^{-\alpha t} d\Omega)^{\frac{1}{2}} (\int_\Omega (\zeta^2 + \zeta_x^2) e^{-\alpha t} d\Omega)^{\frac{1}{2}}$$

$$\leq \sqrt{2} \max\{\lambda_{max}, b_{max}\} \|\eta\|_1 |||\zeta||| =: c_2 \|\eta\|_1 |||\zeta|||$$

because $\sqrt{a} + \sqrt{b} \leq \sqrt{2}\sqrt{a+b}$ for all positive real numbers a and b. These relations above in combination with $|||\theta||| = |||\eta - \zeta||| \leq |||\eta||| + |||\zeta||| \leq \|\eta\|_1 + |||\zeta|||$ lead to

$$|||\theta||| \leq (1 + \frac{c_2}{c_1}) \|\eta\|_1 \, .$$

Finally for $u \in H^{s+1}(\Omega)$ the classical interpolation error estimate

$$\|\eta\|_r \leq D h^{s+1-r} \|u\|_{s+1} \quad \forall_{0 \leq r \leq s \leq 1}$$

gives the discretization error estimate

$$|||\hat{u} - \hat{u}_h||| \leq D(1 + \frac{c_2}{c_1}) h^s \|u\|_{s+1} \quad \forall_{0 \leq s \leq 1} \tag{4.2}$$

which is uniform in ε if $b_{min} > 0$. Since the $|||\cdot|||$ norm is slightly weaker than the $\|\cdot\|_1$ norm, the discretization error estimate is not of optimal order, even if $u \in H^2(\Omega)$, the Sobolev space of order 2.

To analyse the stability of the time-stepping for (2.2) consider again the variational formulation (3.2). Let a perturbation $\delta \in L^2((0,1))$ be defined by $\delta = \delta(x)$ on $(0,1)$, and assume that $\hat{u}$ and $\hat{u}_\delta$ are solutions of $F(u) = 0$ satifying the Dirichlet boundary conditions on $\Gamma_2 \cup \Gamma_4$ with $\hat{u}_{|\Gamma_1} = u_0$ and $\hat{u}_{\delta|\Gamma_1} = u_0 + \delta$.

Note that $\theta := \hat{u} - \hat{u}_\delta \in H^1(\Omega)$ satisfies $\theta \equiv 0$ at $\Gamma_{2,4}$, whence under the appropriate assumptions, posed in the beginning of the section,

$$0 = < F(\hat{u}) - F(\hat{u}_\delta), \hat{u} - \hat{u}_\delta >$$

$$= \int_\Omega \left[(A(\hat{u})\underline{\nabla}\hat{u})^t \underline{\nabla}\theta + (\underline{b}^t \underline{\nabla}\hat{u} + f)\theta \right] e^{-\alpha t} - \left[(A(\hat{u}_\delta)\underline{\nabla}\hat{u}_\delta)^t \underline{\nabla}\theta + (\underline{b}^t \underline{\nabla}\hat{u}_\delta + f)\theta \right] e^{-\alpha t} d\Omega$$

$$= \int_0^1 \int_\Omega \left[(B(\hat{u}_\delta + \tau\theta)\underline{\nabla}\theta)^t \underline{\nabla}\theta \right] e^{-\alpha t} d\Omega \, d\tau + \int_\Omega (\underline{b}^t \underline{\nabla}\theta)\theta e^{-\alpha t} d\Omega$$

$$> \frac{1}{2} \int_\Omega \theta^2 (b_2 \alpha - \nabla \cdot \underline{b}) e^{-\alpha t} d\Omega + \frac{1}{2} \oint_{\partial\Omega} \theta^2 e^{-\alpha t} \underline{b}^t \vec{n} \, ds$$

$$> \frac{1}{2} e^{-\alpha T} \oint_{\Gamma_3} \theta^2 b_2 \, ds - \frac{1}{2} \oint_{\Gamma_1} \theta^2 b_2 \, ds \, .$$

This leads according to the remark below (3.2) to the identity

$$\oint_{\Gamma_3} \sigma(.,t_0+T)(\hat{u}(.,t_0+T) - \hat{u}_\delta(.,t_0+T))^2 \, ds < e^{\alpha T} \oint_{\Gamma_1} \sigma(.,t_0)(\hat{u}(.,t_0) - \hat{u}_\delta(.,t_0))^2 \, ds$$

for all time-slabs.

The above result shows stability in L^2-norm on Γ_1 of initial values for $\alpha = 0$ and $b_2 = \sigma > 0$, a constant, for each time-slab. Hence the repeated 'time-slabbing' can not cause increasing errors in L^2-norm with more than a factor $e^{\alpha sT}$, where s is the total number of time-slabs. By assumption, sT is bounded, independent of any mesh parameter. If $\alpha = 0$, then we have stability, even without this assumption.

In order to investigate the conditions in (4.1) consider, as an example, the diffusion (see also [9])

$$\varepsilon(\zeta) = \varepsilon_{\min} + \varepsilon_{\max} \frac{\arctan(\xi(\zeta - \zeta_0)) + \arctan(\xi\zeta_0)}{\frac{\pi}{2} + \arctan(\xi\zeta_0)}$$

which models the electromagnetic field reluctivity of an electromagnetic field penetrating into a halfspace of ferromagnetic material for appropriately chosen boundary conditions and positive constants $\varepsilon_{\min} \approx 10^{-3} \varepsilon_{\max}$, ξ and ζ_0. In this case $\varepsilon' \geq 0$ and $\lambda(\zeta) := \varepsilon(\zeta) + 2\zeta \varepsilon'(\zeta)$ is a continuous function, bounded above and below on $[0, \infty)$ by

$$\begin{cases} 0 < \lambda_{\min} := \varepsilon_{\min} \leq \lambda(\zeta) \leq \lambda(\zeta_{\max}) =: \lambda_{\max} < \infty & \forall_{\zeta \in [0,\infty)}, \\ \zeta_{\max} = \frac{1}{\xi}((4\xi^2\zeta_0^2 + 3)^{\frac{1}{2}} - \xi\zeta_0) \end{cases}$$

whence the first two conditions of (4.1) are satisfied. Due to $\underline{b} = \begin{bmatrix} 0 \\ 1 \end{bmatrix}$ we have $b_{\min} = 0$ and $b_{\max} = 1 < \infty$, whence all conditions of (4.1) are satisfied for all $\alpha \geq 0$.

5. Numerical results and mesh generation

On each time-slab we first generate a uniform coarse finite element triangulation in the standard way (for a three-dimensional problem we can use tetrahedron elements, see [20]). In the case of a moving shock in time with a known direction or a parabolic layer, the grid will be refined locally to better fit the solution. If there is no information about the solution the mesh will be refined uniformly. However, we advise to use adaptive mesh refinement techniques.

As a refinement strategy we use subsequent refinement of all triangles overlapping a certain curve ('**curve**' technique) or subsequent refinement of all triangles inside a certain area (called '**area**' technique). Because each refinement halves the heights of the triangles involved the ratio of maximal and minimal height ('**Ratio**') is multiplied by a factor 2.

Depending on the choice of the flow field component b_2 there may appear a parabolic layer along Γ_2. If the mesh is not refined here, oscillations would arise with our discretization method even if we use a standard streamline-upwind method, because no artificial diffusion perpendicular to the streamlines is used. However, the use of a fine mesh along this layer makes artificial diffusion unnecessary, and in addition provides an accurate resolution of the layers.

The subsequently approximated parts of the solution are piecewise linear. The old mesh points at Γ_3 of Ω_{t_0} must be used as mesh points for the boundary Γ_1 of the new time-slab Ω_{t_0+T} because otherwise the restriction of the discrete solution on Γ_3 (Ω_{t_0}) will not exactly be represented by the finite element functions on the new subdomain, which might cause additional errors. We may however add more mesh points, whereby Dirichlet boundary conditions are determined by linear interpolation, or we may adjust the mesh so that we end up with fewer points on Γ_3 than on Γ_1 (which is convenient if the solution gets smoother with increasing time).

As an initial solution for the non-linear iterations on each time-slab the initial solution used for the first time-slab is taken, with the Dirichlet boundary condition on Γ_1 determined by the solution on the previous time-slab.

Table 5.1 Mesh generation details

Mesh	Ω	T	Slab	T	N	Ratio	Time	Fig.
1(area)	$(0,1) \times (0,\infty)$	$\frac{1}{4}$	coarse	8	3	1	0.00	3
			all	318	190	32	0.59	6.1a
2(area)	$(0,1) \times (0,\infty)$	$\frac{1}{4}$	coarse	8	3	1	0.00	3
			all	739	405	32	1.11	6.2a
3(curve)	$(0,1) \times (0,\frac{1}{2})$	$\frac{1}{8}$	coarse	16	7	1	0.00	4
			1	1757	906	64	1.83	6.3a
			2	2133	1097	64	1.91	6.3a
			3	2133	1097	64	1.91	6.3a
			4	2110	1086	64	2.15	6.3a
(area)			5	3527	1837	64	3.41	6.3a

Table 5.1 gives a survey of the meshes to be used for the numerical tests below. For each problem, the first time-slab and time period T, as well as the number of triangles T and the number of mesh points N for the coarse initial grid and all subsequent time-slabs, are given. Further, the refinement technique used, together with the ratio of triangle heights and the mesh generation time in seconds are provided. If for all time-slabs the same refined mesh will be used, then this is denoted by 'all' in the table.

In order to study the performance of the techniques proposed, we consider 3 linear problems, described in table 5.2, and refer to [9], for the non-linear problem described at the end of the previous section. Because the problems are linear, in (3.3) there will be only one iteration necessarily to obtain the solution $\hat{u}$, whence it is satisfactory to take $\tau_k = 1$ and $\varepsilon_{\text{non-linear}} = \varepsilon_{\text{linear}}$. Testcase 1 will be a problem with a divergence free velocity field, testcase 2 considers a linear problem with a velocity field of positive divergence, where $\alpha > 0$ has to be chosen in order to get convergence, and finally testcase 3 shows a problem with a shock moving in time. Note that, according to table 5.2, the initial solution on each time-slab satisfies the boundary conditions and is zero at all mesh points inside the domain.

Example 3 is taken into account to show the curve mesh refinement technique, here the time-slabs differ from one time-slab to the next. The difference between examples 1,2 and 3 is that for $\varepsilon \downarrow 0$ there will appear a parabolic boundary layer along Γ_2 in the former cases, whereas there will appear a moving shock inside the domain in the latter case. Note that this latter test is not covered by the provided theory, because the boundary conditions, described by γ, can not be extended to a function in $H^1(\Omega)$.

Table 5.2 shows the number of linear iterations necessary to solve problems 1 and 3 for each time-slab, with the use of the linear solver GCGLS and streamline-upwind basisfunctions, for $\alpha = 0$. The use of streamline-upwind basisfunctions is somewhat more expensive than the use of standard Galerkin basisfunctions, but it guarantees the convergence of the linear iterations. For small values of ε it is not possible to obtain convergence of the linear solver

Table 5.2 The testcases

Testcase, Mesh	1	2	3
Ω	$(0,1) \cdot (0,\infty)$	$(0,1) \cdot (0,\infty)$	$(0,1) \cdot (0,\frac{1}{2})$
ε	$1 \cdot 10^{-4}$	$1.6 \cdot 10^{-3}$	$1 \cdot 10^{-6}$
α	0	1	0
$\underline{b}$	$[1,1]^t$	$[x,1]^t$	$[2,1]^t$
f	0	0	0
$l(t)$	0	0	0
$r(t)$	$\sin(8\pi t)$	$\sin(8\pi t)$	1
$u_0(x)$ on Γ_{1,Ω_0}	0	0	1
$u^0(x,t)$ on Ω_{t_0+T}	0	0	0
$\varepsilon_{\text{linear}}$	$1 \cdot 10^{-10}$	$1 \cdot 10^{-10}$	$1 \cdot 10^{-10}$
Remarks	Layer at Γ_2	Layer at Γ_2	Moving shock

Table 5.3 Number of iterations for tests 1 and 3

Test\ Slab	1	2	3	4	5
1	14*	14*	14*	14*	14*
3	73*	75*	77*	78*	28*

*: A SUPG discretization used because no or too slow convergence of the linear solver for the FE discretization

if the FE discretization method is used. However, note that there are no reasons to use the streamline-upwind basisfunctions if there are no layers in the solution.

To accelerate the linear solvers an ILU factorization is used as a preconditioner (see for instance [6] and the references quoted therein).

Table 5.4 shows that for test 2 the FE discretization leads to a too slow convergence of the linear solver for small values of ε, whence a SUPG discretization had to be used. Because for this test $-\nabla \cdot \underline{b} < 0$, the weight α has to be greater than zero for small values of ε. The table illustrates that there will be no convergence of the Galerkin finite element discretization for $\alpha < 1$ if $\varepsilon \downarrow 0$. For practical reasons α must not be chosen too large.

Table 5.4 Number of iterations for test 2

$\alpha \backslash \varepsilon \cdot 10^3$	10	1.70	1.65	1.60	1.50	1.00
$\frac{1}{2}$	25	62	20*	20*	21*	19*
1	25	38	40	78	19*	18*

*: A SUPG discretization used because no or too slow convergence of the linear solver for the FE discretization

Figures 3 and 6.1a show the coarse initial mesh resp. refined mesh generated for the first test. Because there will a parabolic boundary layer along Γ_2, the refined mesh is only fine in a small area along this boundary. The equidistant levels of the SUPG solution as well as the SUPG solution itself are shown in figs 6.1b and 6.1c for the first time-slab and fifth time-slab respectively.

Figures 6.2a-6.2c show the refined mesh and compare the FE solution with the SUPG solution, for the second testcase (the initial coarse mesh is the same as for problem 1). Note that this mesh is refined over a larger area and that the FE solution is instable, as is to be expected.

The coarse mesh for the third problem is shown in fig. 4 and the five refined time-slabs, required to solve this problem, are shown in fig. 6.3a. For the first, fourth and fifth time-slab the SUPG solution together with its equidistant contours are given in figs. 6.3b,c and 6.3d respectively. The mesh for each time-slab is constructed as follows. First the position of the shock along Γ_1 is determined. Then the shock curve is determined by following the solution independent velocity field $\underline{b}$, up to Γ_3. Finally the 'curve' refinement technique is used 6 times along this shock curve. Note that the shock is captured well inside the fine mesh elements which are generated along the shock curve. However, in general we would recommend an alternative method based on adaptive refinement.

6. Conclusions

We have demonstrated the efficiency of using finite elements in both time and space where we consider the time-space domain as a whole in the generation of finite elements. The method is applicable also in multi-dimensional problems where we use tetrahedron elements, for instance. As we have seen the stability of time-stepping on the larger timeslabs is an immediate consequence of the positive definiteness of the Hessian. The use of ordinary continuous finite element approximations enables us to use standard finite element packages for the time-space domain. Adaptive refinement of an initial mesh on each time-slab in order to locate and fit steep gradients is advisable and is presently studied by one of the authors.

The solution of the linear systems can be performed quite cheaply. Using still more efficient preconditioners for instance based on incomplete factorization and domain decomposition (see [7] and [10]), we can get methods for which the computational effort is not larger than about proportional to the number of mesh points. This means that we can get savings in the computational effort of orders of magnitude compared to standard time-stepping methods, even if moving mesh strategies are used, when problems with local layers are solved.

Finally, note that we could have alternatively used higher order instead of piecewise linear finite elements with obvious minor modifications done in the above presentation. This would have enabled us to get a faster rate of convergence of the discrete approximations to the solution of the partial differential equation.

7. References

[1] Axelsson,O., *On global convergence of iterative methods,* in Iterative Solution of Nonlinear Systems of Equations, LNM#953 (Ansorge R., Meis Th. and Törnig W., eds.), pp.1-19, Springer Verlag 1982

[2] Axelsson,O., *A generalized conjugate gradient, least square method,* Numer. Math. 51 (1987), 209-227

[3] Axelsson,O., *Finite element methods for convection diffusion problems,* in Numerical Treatment of Differential Equations (Strehmel, ed.), [Proceedings of the Fourth Seminar "NUMDIFF-4", Halle, 1987], Teubner-Texte zur Mathematik, Band 104, Teubner, Leipzig, pp.171-182

[4] Axelsson,O., *The numerical solution of parial differential equations,* in Mathematics and Computer Science II: fundamental contributions in the Netherlands since 1945 (Hazewinkel,M., Lenstra,J.K. and Meertens,L. eds.), pp.1-18, North-Holland 1986

[5] Axelsson,O., *On the numerical solution of convection dominated convection-diffusion problems,* in Mathematical Methods in Energy Research (Gross,K.I. ed.), pp.3-21, SIAM Philadelphia 1984

[6] Axelsson,O. and Barker,V.A., *Finite Element Solution of Boundary Value Problems,* Academic Press, Orlando, Fl., 1984

[7] Axelsson,O., Eijkhout,V., Polman,B. and Vassilevski,P., *Iterative solution of singular pertubation 2nd order boundary value problems by use of incomplete block-factorization*

methods, Intern. rep. 8709, Department of Mathmatics, University of Nijmegen, The Netherlands 1987

[8] Axelsson,O. and Maubach,J., *On the updating and assembly of the hessian matrix in finite element methods,* Computer Methods in Applied Mechanics and Engineering 71(1988), 41-67

[9] Axelsson,O. and Maubach,J., *A time-space finite element discretization technique for the calculation of the electromagnetic field in ferromagnetic materials,* International Journal for Numerical Methods in Engineering, to appear

[10] Axelsson,O. and Polman,B., *A robust preconditioner based on algebraic substructuring and two-level grids,* in Robust Multi-Grid Methods (Hackbusch,W. ed.), Notes on Numerical Fluid Mechanics, Vol. 23, Vieweg, BraunSchweig, 1988, pp.1-26

[11] Axelsson,O. and Steihaug,T., *Some computational aspects in the numerical solution of parabolic equations,* Journal of Computational and Applied Mathematics, 4(1978), 129-142

[12] Axelsson,O. and Verwer,J.G., *Boundary value techniques for initial value problems in ordinary differential equations,* Math. Comp., 45(1985), 153-171

[13] Ciarlet,P.G., *The Finite Element Method for Elliptic Problems,* North-Holland Publ., Amsterdam, 1978

[14] V.d.Houwen,P.J., *Construction of Integration Formulas for Initial Value Problems,* North Holland, Amsterdam 1976

[15] Hughes,T.J. and Brooks,A., *A multidimensional upwind scheme with no crosswind diffusion,* in AMD 34(1979), Finite element methods for convection dominated flows (Hughes, T.J. ed.), ASME, New York

[16] Hughes,J.R. and Hulbert,M., *Space-time finite element methods for elastodynamics: formulation and error estimates,* Computer Methods in Applied Mechanics and Engineering 66(1988), 339-363

[17] Jamet,P., *Galerkin-type approximations which are discontinuous in time for parabolic equations in a variable domain,* SIAM J. Num. Anal. 15(1978), 912-928

[18] Johnson,C. and Pitkäranta,J., *An analysis of the discontinuous Galerkin method for a scalar hyperbolic equation,* Rep. MAT-A215, Institute of Mathematics, Helsinki University of Technology, Helsinki, Finland 1984

[19] Karlsson,K-E. and Wolfbrandt,A., *An explicit technique for calculating the electromagnetic field and power losses in ferromagnetic materials,* Intern. rep. 721-83, department for electrical analysis methods ASEA, Västerås, Sweden, 1983

[20] Lesaint,P. and Raviart,P.A., *On a finite element method for solving the neutron transport equation,* in Mathematical Aspects of Finite Elements in Partial Differential Equations (de Boor,C. ed.), Academic Press, New York 1974, pp.89-123

[21] Nečas,J., *Les Méthodes Directes en Théorie des Equations Elliptiques,* Masson, Paris, 1967

[22] Sonneveld,P., *CGS, a fast Lanczos-type solver for non-symmetric linear systems,* SIAM J. Sci. Stat. Comput. 10(1989), 36-52

[23] Zienkiewicz,O., *The Finite Element Method in Engineering Science,* 3rd edition, Mc Graw-Hill, New York, 1977

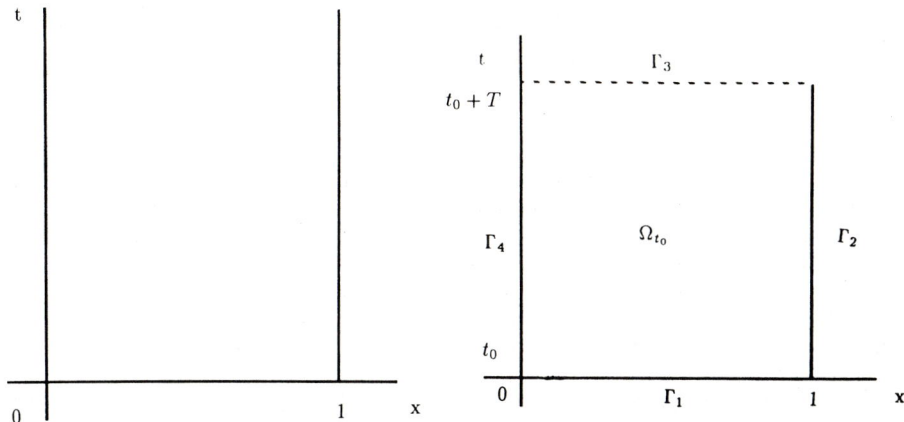

Fig. 1 Time-space domain

Fig. 2 Time-slab Ω_{t_0}

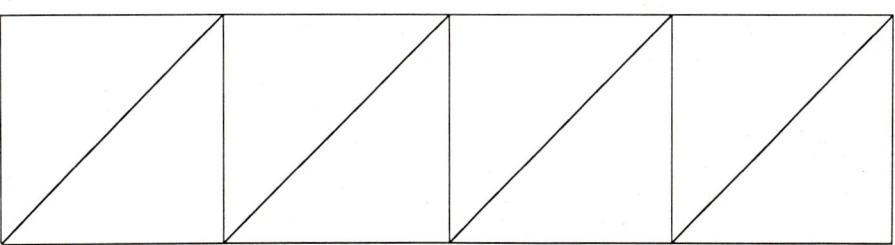

Fig. 3 Coarse initial mesh for tests 1 and 2

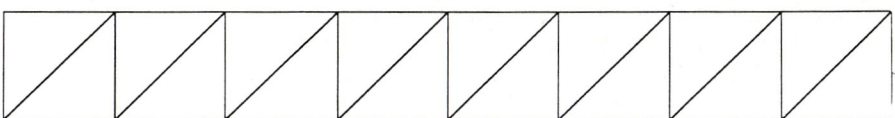

Fig. 4 Coarse initial mesh for test 3

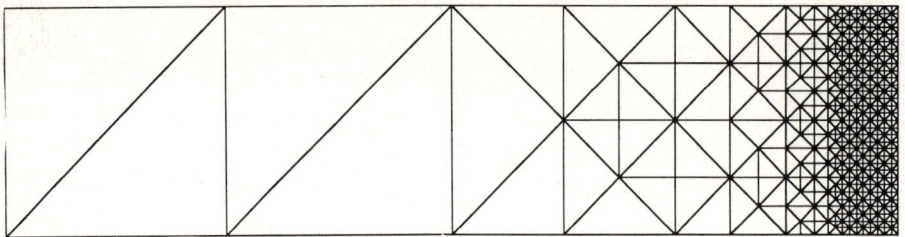

Fig. 6.1a Test 1. Refined mesh for all time-slabs

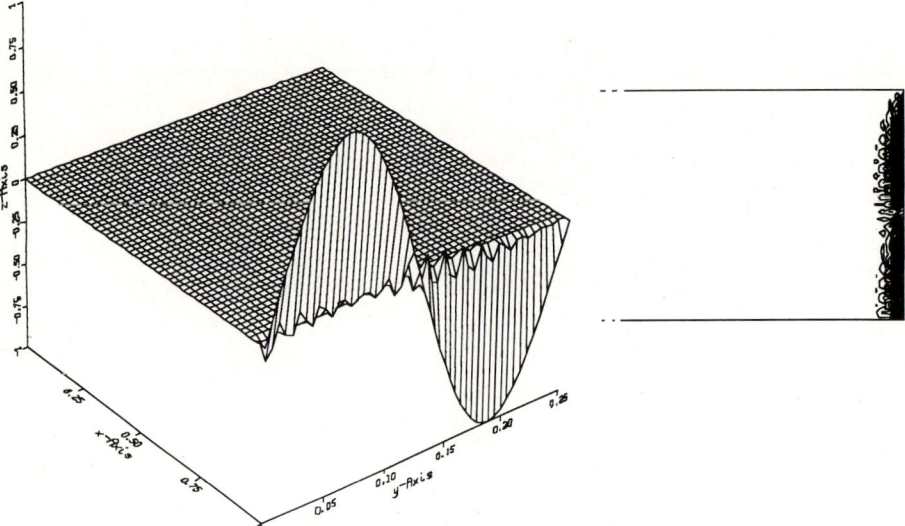

Fig. 6.1b Test 1. Isoclines of and SUPG solution at time-slab 1

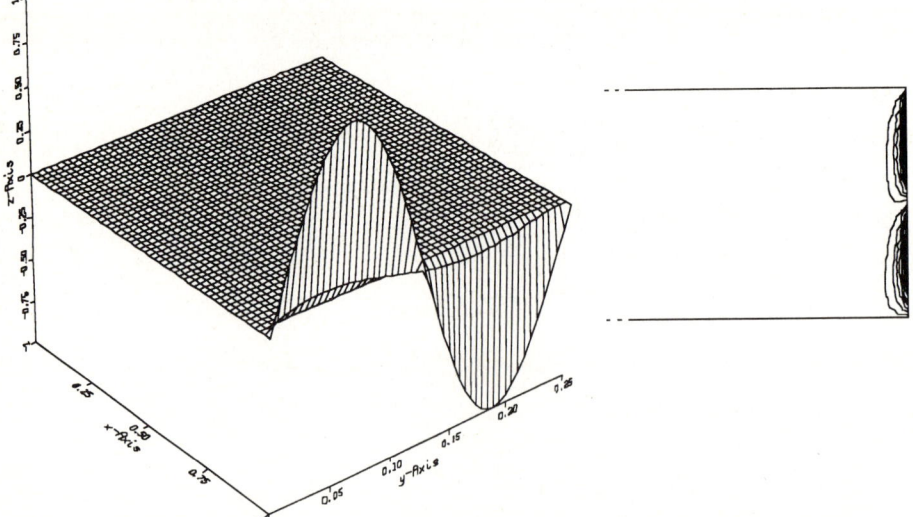

Fig. 6.1c Test 1. Isoclines of and SUPG solution at time-slab 5

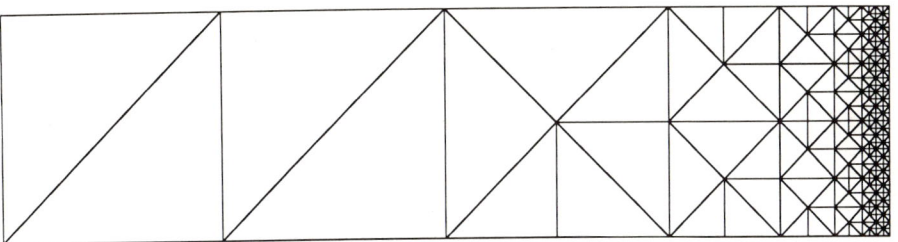

Fig. 6.2a Test 2. Refined mesh for all time-slabs

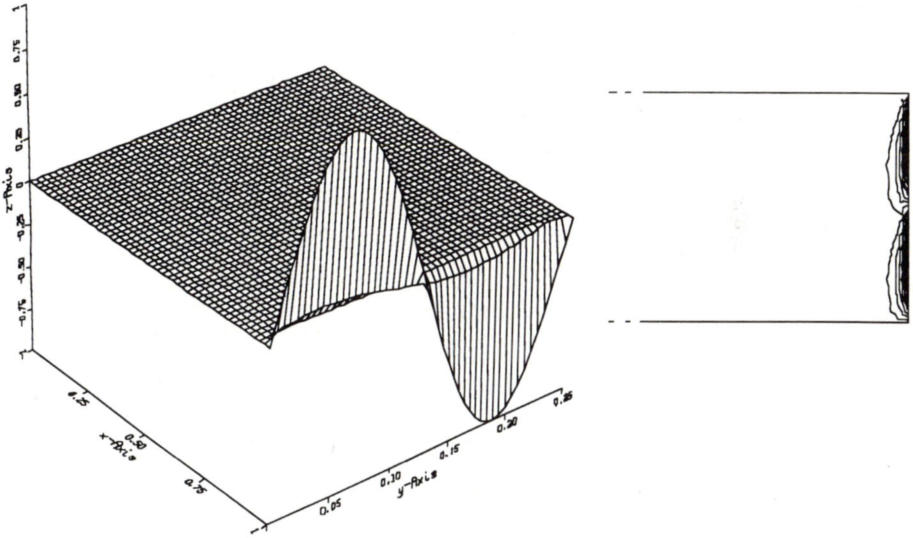

Fig. 6.2b Test 2. Isoclines of and FE solution at time-slab 1

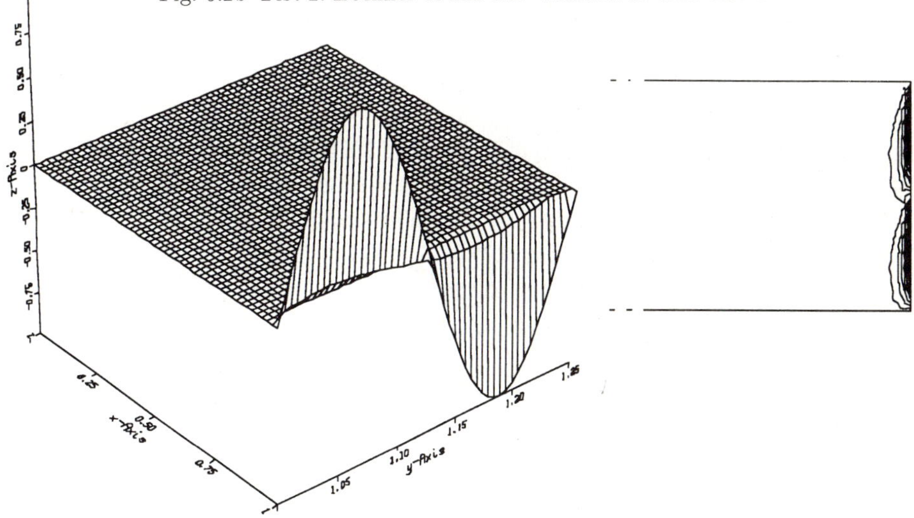

Fig. 6.2c Test 2. Isoclines of and SUPG solution at time-slab 1

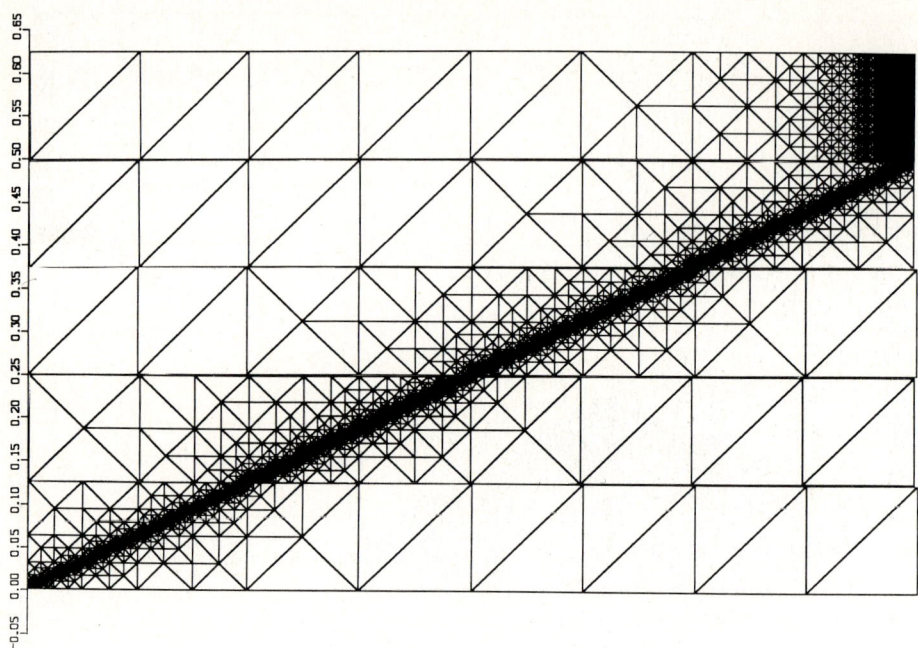

Fig. 6.3a Test 3. Chain of refined meshes for time-slabs 1-5

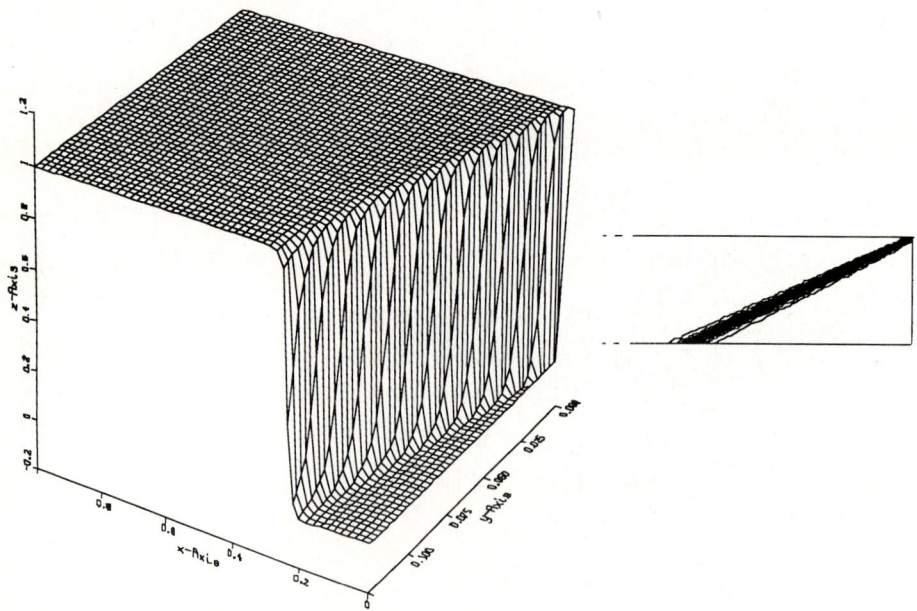

Fig. 6.3b Test 3. Isoclines of and SUPG solution at time-slab 1

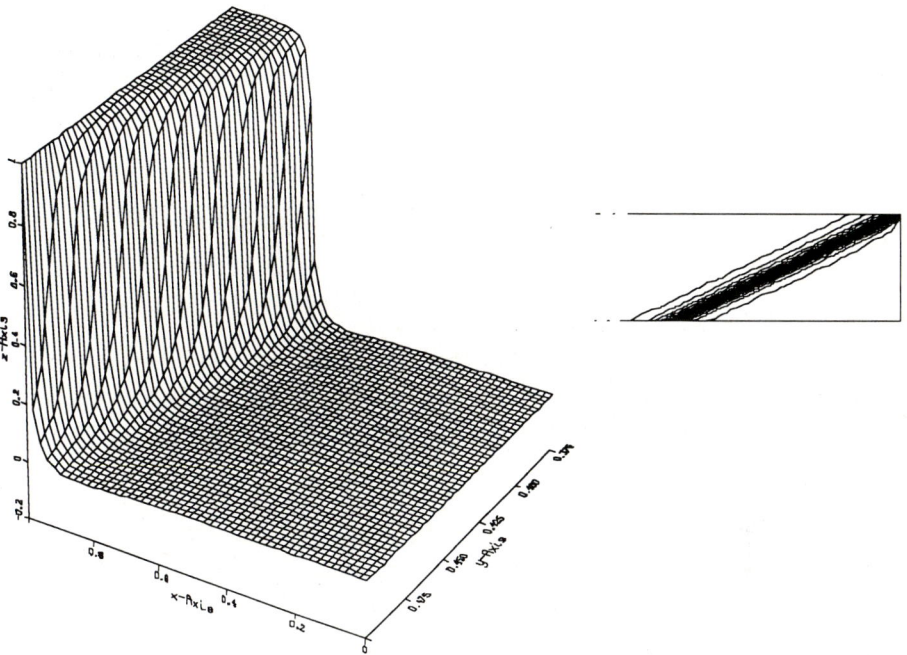

Fig. 6.3c Test 3. Isoclines of and SUPG solution at time-slab 4

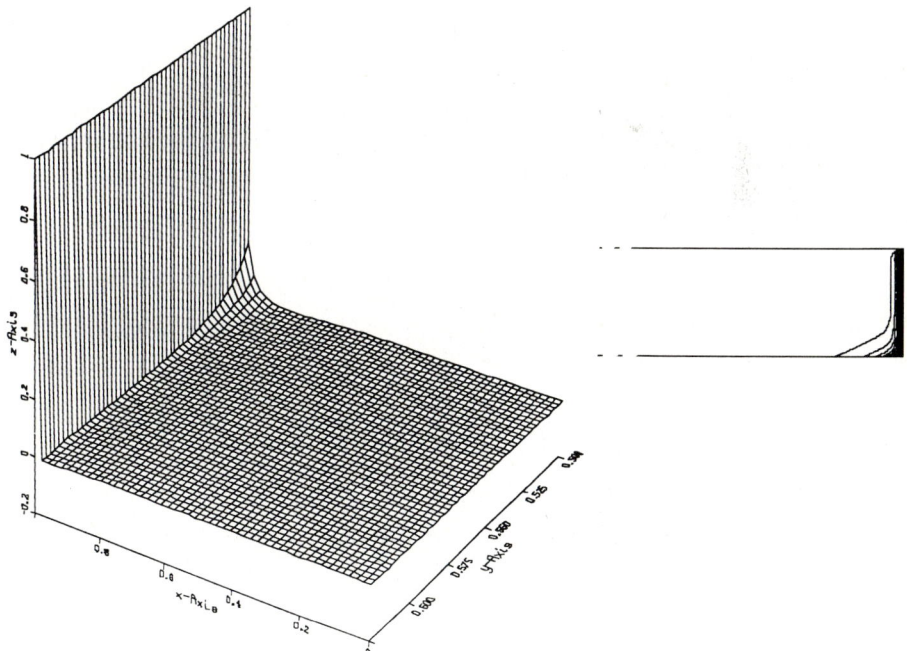

Fig. 6.3d Test 3. Isoclines of and SUPG solution at time-slab 5

Parallelization of Robust Multi-Grid Methods: ILU Factorization and Frequency Decomposition Method

Peter Bastian and Graham Horton [1]

Institut für Mathematische Maschinen und Datenverarbeitung III,
Universität Erlangen-Nürnberg, Martensstraße 3, D-8520 Erlangen

Summary

Using the anisotropic equation as a test problem, the concept of robustness is defined. Two multi-grid methods which are known to have this property are described: the standard multi-grid algorithm with ILU smoothing, and the frequency decomposition method. The parallelization on a MIMD computer is presented, together with results for the speedup obtained. The methods are compared with a standard parallel multi-grid algorithm using a Gauß-Seidel red-black smoother.

1. Introduction

The numerical solution of the Navier-Stokes equations with multi-grid methods has been investigated by several authors (see [2],[8]). Two problems occuring in practice are convection dominated flows and complex geometries requiring the use of body-fitted grids. The influence of these problems on the rate of convergence can be exemplified by two scalar test equations: the *anisotropic* equation

$$-\alpha \frac{\partial^2}{\partial x^2} u(x,y) - \beta \frac{\partial^2}{\partial y^2} u(x,y) = f(x,y) \qquad (1)$$

with $\Omega = [0,1]^2$, $u(x,y) = b(x,y)$ on $\partial\Omega$, $\alpha, \beta > 0$, and the *convection-diffusion* equation

$$-\epsilon \left(\frac{\partial^2}{\partial x^2} u(x,y) + \frac{\partial^2}{\partial y^2} u(x,y) \right) + c_1 \frac{\partial}{\partial x} u(x,y) + c_2 \frac{\partial}{\partial y} u(x,y) = f(x,y) \qquad (2)$$

with $\Omega = [0,1]^2$, $u(x,y) = b(x,y)$ on $\partial\Omega$ and $\epsilon > 0$.

A standard discretization of Eqn. (1) with finite differences on a grid with meshsize h_l is

$$A_l = h_l^{-2} \begin{bmatrix} & -\beta & \\ -\alpha & 2(\alpha+\beta) & -\alpha \\ & -\beta & \end{bmatrix}, \quad h_l = \frac{1}{N_l}, N_l = 2^{l+1} \qquad (3)$$

---

[1] The authors were supported in part by the Stiftung Volkswagenwerk within its program 'Entwicklung von Berechnungsverfahren für Probleme der Strömungstechnik'

and $\alpha \ll \beta$ or $\alpha \gg \beta$ models a discretization of Poisson's equation on a grid with high aspect ratio. The discretization (3) leads to a system of linear equations

$$A_l u_l = f_l \ . \tag{4}$$

If $K_l(\alpha, \beta)$ is the iteration matrix of an iterative method for the solution of the discrete problem Eqn. (4), then the method $K_l(\alpha, \beta)$ is called *robust for the anisotropic equation* iff

$$\|K_l(\alpha, \beta)\| \leq \zeta < 1 \ \forall \alpha, \beta > 0, \ l \in \mathbf{N} \tag{5}$$

with $\|.\|$ the spectral norm. The robustness of a method for a discretization of Eqn. (2) can be defined similarly.

The definition (5) does not imply how to achieve robustness. Assuming $K_l(\alpha, \beta)$ to be the iteration matrix of a standard multi-grid algorithm, the error in the n-th iterate $v_l^n = u_l - u_l^n$ is given by $v_l^n = K_l(\alpha, \beta) v_l^{n-1}$. With

$$(e_l^{\nu\mu})_{i,j} = \sin(\nu \pi i h_l) \sin(\mu \pi j h_l) \tag{6}$$

the eigenvectors of A_l, one defines the *low frequencies* $V_{00} = \text{span}\{e_l^{\nu\mu} | \nu, \mu \in \mathcal{L}\}$, the *high frequencies in x-direction* $V_{10} = \text{span}\{e_l^{\nu\mu} | \nu \in \mathcal{H}, \mu \in \mathcal{L}\}$, the *high frequencies in y-direction* $V_{01} = \text{span}\{e_l^{\nu\mu} | \nu \in \mathcal{L}, \mu \in \mathcal{H}\}$ and the *high frequencies in both directions* $V_{11} = \text{span}\{e_l^{\nu\mu} | \nu, \mu \in \mathcal{H}\}$ with $\mathcal{L} = \{1, \ldots, N_{l-1} - 1\}$ and $\mathcal{H} = \{N_{l-1}, \ldots, N_l - 1\}$. The basic idea behind the multi-grid method is that the smoother reduces the errors $v \in V_{10} \cup V_{01} \cup V_{11}$ (the *high frequencies*) and that the coarse grid correction reduces the errors $v \in V_{00}$. Fourier analysis of the two-grid method with a damped Jacobi smoother applied to Eqn. (4) yields that the smoother fails to reduce the errors in V_{10} (V_{01}) if $\alpha \to 0$ ($\beta \to 0$). This leads to a first way to achieve robustness: incorporating a more complex smoothing algorithm that works for all high frequencies. Wittum has shown in [7] that the multi-grid algorithm with ILU_β smoother is indeed robust for the anisotropic equation. Another approach is the frequency decomposition method, where four different coarse grid corrections are constructed for each of the spaces V_ι, $\iota \in \{00, 10, 01, 11\}$. The frequency decomposition method has been developed by Hackbusch and is described for periodic boundary conditions in [4].

In the next section the principles of parallel multi-grid algorithms will be shortly reviewed. Sections three and four describe the ILU iteration and the frequency decomposition method, respectively, together with the parallelization and speedup results for each method. In the last section the two methods will be compared with a parallel multi-grid method using a highly parallelizable red-black Gauß-Seidel smoother.

2. Parallelization of Multi-Grid Methods

2.1 General Remarks

If we assume that a two-dimensional partial differential equation is discretized on a rectangular grid with $(N_l - 1) \times (N_l - 1)$ grid points then each processor is assigned to a subset of the unknowns. In a one-dimensional arrangement of n processors called a *ring* configuration of length n, each processor is assigned to $((N_l-1)/n)\times(N_l-1)$ unknowns. In a $n\times m$ array configuration each processor is assigned to $((N_l-1)/n)\times((N_l-1)/m)$ unknowns.

All components of the multi-grid method require only *local* operations, i.e. computations at grid point (i,j) need only values at the grid points $(i\pm 1, j\pm 1)$. In addition most operations (defect calculation, restriction,...) can be carried out in parallel at each point. The number of points that can be computed simultaneously varies however between different smoothing procedures: This number is $(N_l-1)^2$ for Jacobi smoothing, $(N_l-1)^2/2$ for red-black smoothing and at most N_l-1 for ILU and lexicographic Gauß-Seidel smoothers.

The *speedup* $S(p)$ of a parallel algorithm is defined as

$$S(p) = \frac{T_{Mono}}{T_{Multi}(p)} \qquad (7)$$

with T_{Mono} the time needed for the execution of the program on a single computer and $T_{Multi}(p)$ the time taken on p processors. This is the factor by which computation time is reduced through the use of a parallel processor. Sometimes we refer to the *efficiency* of a parallel implementation which is defined as

$$E(p) = \frac{S(p)}{p} \ . \qquad (8)$$

The speedup for all operations carried out on a fixed level l within the multi-grid method depends heavily on the number of unknowns per processor, which means that a high speedup can be achieved on the fine grids (assuming a large number of grid points per processor on the fine grids), whereas the speedup deteriorates on the coarser grids. Nevertheless the overall speedup can be expected to be very high because most of the work is spent on the finest levels.

2.2 The DIRMU System

DIRMU is a multiprocessor consisting of 25 stand-alone micro-computers developed at the University of Erlangen-Nürnberg [5]. The processors communicate via *distributed shared* memories whereby each processor can be connected to the memories of up to seven other processors. In addition to this communication memory, called *multiport memory*, each processor has a private memory where the program and local data are stored. All variables shared with neighbouring processors are stored in the multiport memories in order to avoid an explicit data transport. The distributed operating system DIRMOS supplies information about various possible connection structures like ring, array, tree or hypercube to a parallel program. In this way it is possible to use a connection structure that best fits the problem. DIRMOS also allows multiple users to run parallel programs on different parts of the system simultaneously and independently of each other.

3. The Incomplete LU Factorization

3.1 The Algorithm

The Incomplete Decomposition Method (ILU) has found widespread application in Navier-Stokes solvers, both as a solution procedure in its own right, and as a preconditioner for CG methods or as a smoother in multi-grid methods ([8],[2]). In addition to the basic method, several variants have been introduced, such as the Semi-Implicit Procedure ([9]), or more recently ILU_β ([7]). In addition the method is applicable to 5, 7, and 9-point discretization molecules for two-dimensional problems. In this paper, the parallelization of the simplest 5-point version will be described for simplicity, although the technique is easily extended to the more complex methods.

In Eqn. (4) the matrix A_l is pentadiagonal if the grid points are ordered lexicographically first in x then in y-direction. The discretization molecule (3) has constant coefficients but in general we allow variable coefficients denoted by (we will skip the index l in the rest of this section):

$$A_l(i,j) = \begin{bmatrix} & A^N_{i,j} & \\ A^W_{i,j} & A^C_{i,j} & A^E_{i,j} \\ & A^S_{i,j} & \end{bmatrix}. \qquad (9)$$

ILU is an iterative method which uses an approximate factorization of A into a product of upper and lower triangular matrices L and U with prescribed sparsity pattern, whereby the coefficients of the matrices L and U have the same notation as the corresponding elements of the matrix A.

The matrix A is replaced by $LU - C$, where C is the error in the approximation. This yields the following iteration, which is used as a smoother in a standard multi-grid algorithm:

$$u^{n+1} = u^n + U^{-1}L^{-1}(f - Au^n). \qquad (10)$$

The entries of L and U are computed from the entries of A in the following manner:

$$\begin{aligned} L^S_{i,j} &= A^S_{i,j}, \quad L^W_{i,j} = A^W_{i,j} \\ L^C_{i,j} &= A^C_{i,j} - \frac{A^W_{i,j} A^E_{i-1,j}}{L^C_{i-1,j}} - \frac{A^S_{i,j} A^N_{i,j-1}}{L^C_{i,j-1}} \\ U^C_{i,j} &= 1, \quad U^E_{i,j} = \frac{A^E_{i,j}}{L^C_{i,j}}, \quad U^N_{i,j} = \frac{A^N_{i,j}}{L^C_{i,j}}. \end{aligned} \qquad (11)$$

The computation of the coefficients $L^C_{i,j}$ is recursive, as these depend on the corresponding coefficients in neighbouring points $L^C_{i-1,j}$ and $L^C_{i,j-1}$. Such recursions are in general non-parallelizable.

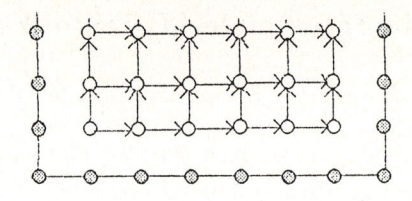

Figure 1: Data Dependencies for the ILU Decomposition

3.2 Parallelization

The key to the parallelization lies in the observation that when Dirichlet or von Neumann boundary conditions are given, the 'west' coefficients of the leftmost variables $A_{1,j}^W$ and the 'east' coefficients of the rightmost variables $A_{N-1,j}^E$ are equal to 0. This means that the term $A_{i,j}^W A_{i-1,j}^E (L_{i-1,j}^C)^{-1}$ drops out of Eqn. (11) and that therefore $L_{1,j}^C$ depends only on $L_{1,j-1}^C$. In Figure 1 an arrow from point (i,j) to grid point (i',j') indicates that grid point (i,j) has to be computed before grid point (i',j').

Two-dimensional grids are mapped onto a ring of processors as described in section two. Grids which are to be accessed by neighbouring processors are located in the multiport memory, the others in private memory. The following algorithm, which is executed simultaneously by each processor, shows the parallel computation of the coefficients $L_{i,j}^C$, where the variables $xmin(p)$ and $xmax(p)$ contain the leftmost and rightmost grid point column stored by processor p. Here the execution of routine signal_row(j) by processors p indicates the completion of its segment of row j by incrementing a counter in the multiport memory of its right neighbour. The procedure wait_row(j) waits for this signal in order to be able to proceed with the computation of row j.

Algorithm 1 *Parallel computation of the L^C entries.*

```
PROCEDURE computeLC (A, L);
BEGIN
   FOR j := 1 TO N − 1 DO
      wait_row(j);
      FOR i := xmin(p) TO xmax(p) DO
```
$$L_{i,j}^C = A_{i,j}^C - A_{i,j}^W (L_{i-1,j}^C)^{-1} A_{i-1,j}^E - A_{i,j}^S (L_{i,j-1}^C)^{-1} A_{i,j-1}^N;$$
```
      END;
      signal_row(j);
   END;
END computeLC ;
```

Zero entries in A^W and A^E are reproduced in L^W and U^E, so that the solution of the systems L and U is parallelizable in an analogue fashion.

The speedup of the parallel ILU procedure can be predicted by a theoretical model. This was done for several reasons.

- A detailed understanding of the behaviour of the parallel method is helpful for optimization purposes.

- The model can be used to predict speedups for large problems whose memory requirements exceed the capacity of the DIRMU system.

- Estimated speedups for larger processor configurations can be obtained.

- The influence of the multiprocessor architecture can be examined by variation of the model parameters.

Here a simplified version of the model is given:

$$S(p, N) = = \frac{(N-1)^2 * T_{OP}}{T_{startup} + T_{parallel}} \tag{12}$$

$$T_{startup} = (p-1) * T_{SY} + (N - \frac{N}{p}) * T_{OP} \tag{13}$$

$$T_{parallel} = (N-2) * T_{SY} + (N-1) * \frac{N}{p} * T_{OP} . \tag{14}$$

$T_{startup}$ is the time the rightmost processor must wait until it can begin its computation. $T_{parallel}$ is then the time taken until the parallel computations are complete. T_{SY} (Synchronization time) is the time needed for a **signal_row** and **wait_row** pair, and T_{OP} (Operation time) is the time needed for the arithmetical computations at a single grid point. As can be seen from the model, the ratio T_{SY}/T_{OP} will be an important factor in the resulting speedup. Note that with DIRMU, all the data required by each processor is directly available in its own memory or in a neighbouring multiport memory, which makes T_{SY} small. In a message-passing parallel computer, T_{SY} will be the sum of the message setup time and the time required to transport the value of a boundary point $L^C_{xmax(p),j}$ from one processor to the next. Fig. 2 shows a comparison of the measured efficiency of the ILU method and the predictions obtained by an optimized model for various configuration lengths and grid sizes. Fig. 3 shows the predictions obtained for larger problems and configurations. Finally, Table 1 shows the efficiency of the multigrid method with ILU smoothing and $h = 1/64$ on the finest grid.

The model is found to predict the measured efficiency very well for the grid and configuration sizes available on DIRMU. It is seen that $h > 1/64$ gives 'reasonable' grid sizes for DIRMU, with efficiencies over 75% for up to 14 processors for $h = 1/64$, and up to 23 processors for $h = 1/128$. The results are typical for a parallel algorithm with data-distribution:

- For a given problem size the efficiency sinks when the number of processors is increased, owing to the growing communication requirements relative to the number of parallelizable operations.

- Large grids can be more efficiently processed than smaller ones

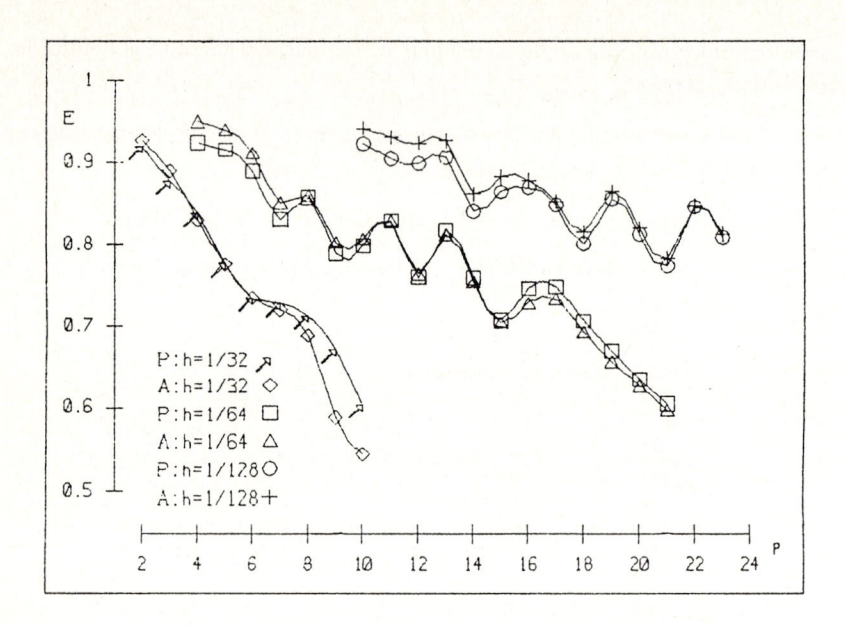

Figure 2: Actual (A) and Predicted (P) ILU Efficiencies

Table 1: Efficiency for multi-grid method with ILU smoothing. Multi-grid data: $\nu_1 = 1$, $\nu_2 = 0$, $\gamma = 1$, $h = 1/64$

p	5	6	7	8	9	10	11	12	13	14	15	16
$E(p)$	0.88	0.82	0.76	0.73	0.67	0.64	0.65	0.58	0.54	0.50	0.46	0.44

- Unequal distribution of data leads to marked 'wiggles' in the curves. The most efficient computations are achieved when the load is evenly balanced among all processors.

The model was found to be accurate enough to enable predictions for larger problems. Here efficiencies better than 75% can be obtained with up to 64 processors for a grid of side length 256, and the efficiency is better than 90% for grids with side length 512 or larger.

4. The Frequency Decomposition Method

4.1 The Algorithm

As pointed out in the introduction, the frequency decomposition method uses a different coarse grid correction for each of the four frequency quadrants. The new

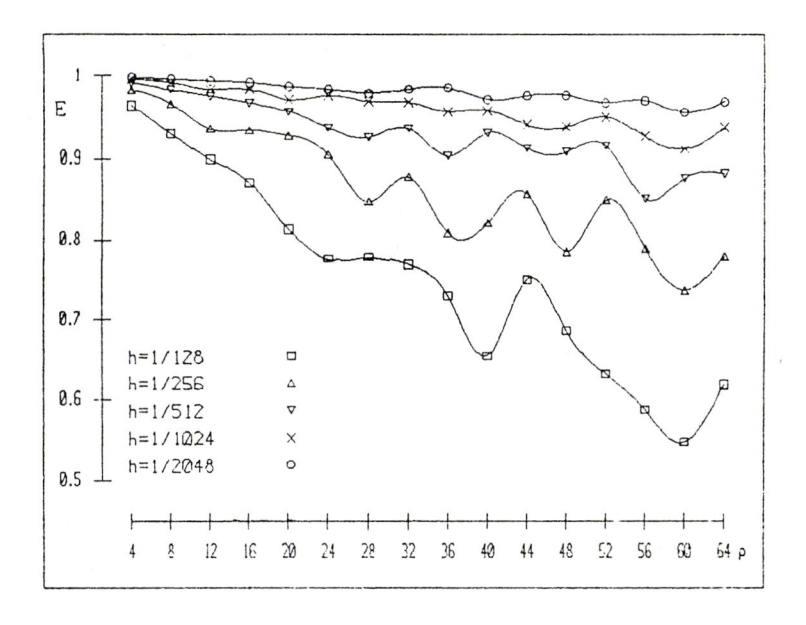

Figure 3: Predicted ILU efficiencies for large configurations

prolongations p_ι and restrictions r_ι ($\iota \in \{00, 10, 01, 11\}$) enable a representation of high frequency error functions on coarser grids, where the coarse grid matrices are computed via the Galerkin product: $A_{l-1}^\iota = r_\iota A_l p_\iota$. This gives us the following algorithm:

Algorithm 2 *The frequency decomposition multi-grid algorithm.*

```
PROCEDURE fdm (l, A_l, u_l, f_l);
BEGIN
    IF l = 0 THEN u_l := A_l^{-1} f_l
    ELSE
        u_l := S_l^{ν1}(u_l, f_l);                          (* pre-smoothing *)
        d_l := f_l - A_l u_l; v_l := 0;                    (* coarse grid correction *)
        FOR ι ∈ {00, 10, 01, 11} DO
            d_{l-1}^ι := r_ι d_l;
            v_{l-1}^ι := 0;
            FOR k := 1 TO γ DO fdm(l - 1, r_ι A_l p_ι, v_{l-1}^ι, d_{l-1}^ι) END;
            v_l := v_l + p_ι v_{l-1}^ι;
        END;
        u_l := u_l + v_l;
        u_l := S_l^{ν2}(u_l, f_l);                          (* post-smoothing *)
    END;
END fdm ;
```

As the algorithm shows, the four corrections can be computed independently of each other. The four matrices A_{l-1}^ι on level $l-1$ produce 16 coarse grid matrices $A_{l-1}^{\iota\kappa}$ on level $l-2$ which means in general that 4^k coarse grid equations result at level $l-k$.

Hackbusch showed that not all of these coarse grid equations are necessary and gave a criterion that results in the *necessary coarse grid tree* of Figure 4. Note that most of the nodes have two descendants (type II nodes in the following) and only some nodes have four descendants (type I nodes). The number of coarse grid equations on level $l-k$ is thus reduced to $3 \cdot 2^k - 2$. For further details of the method we refer to [4].

4.2 Parallelization

The frequency decomposition multi-grid algorithm will be highly parallelizable because firstly Jacobi or red-black Gauß-Seidel smoothers are sufficient, and secondly because the multiple coarse grid corrections can be computed in parallel.

This is exploited by using a two-level strategy. The fine grids, where the number of grid points per processor is assumed to be large, are distributed amongst all processors of the configuration as described in section 2. On the coarser grids the configuration is split into parts and each subconfiguration is assigned to a different coarse grid correction, thus increasing the number of unknowns per processor. The level where the subdivision process starts can be prescribed by the parameter l_{subdiv}.

The subdivision strategy is as follows: If a $n \times 2m$ array configuration is used on level l and a type II node is encountered, then the configuration is subdivided into two equal halves of size $n \times m$. Note that the whole defect grid function is needed in each of the two halves, which requires an explicit data transport, and that the situation is reversed when the two corrections have been computed. If a type I node is encountered, the configuration is also split into two equal halves, where first the 00 and 10 corrections are computed in parallel followed by the parallel computation of the 01 and 11 corrections. Now the strategy is applied recursively, whereby subconfigurations can be further divided.

The parallel computation of the 00 and 10 corrections, each on one half of the configuration, results in an unequal load balance, as can be seen from Figure 4, where the dashed subtree has more nodes than the dotted subtree. This can be partially balanced by using a different parameter γ in the multi-grid cycle for type I and type II nodes. If $\gamma = 1$ is used when calling a type I node, and $\gamma = 2$ is used when calling a type II node, the work spent in the two subtrees is almost equal.

In order to make the exact solver on level 0 as efficient as possible (in the case of periodic boundary conditions level 0 grids have four unknowns) the level 0 grids should be located in a single processor. To achieve this, a configuration of the form $2^{n'} \times 2^{m'}$ has to be used on the finest level. Now, after $n' + m'$ subdividing steps, $2^{n'+m'}$ configurations of size 1×1 are reached. This restriction results in a lower bound for the parameter l_{subdiv}: $l_{subdiv} \geq n' + m'$.

The solution of a p.d.e. with periodic boundary conditions requires the use of a *torus* connection structure. In addition, the subconfigurations arising in the recursive subdivision process must also have torus connections. It turns out that the hypercube structure is the optimal configuration for the two-dimensional frequency decomposition multi-grid method:

- A torus of the form $2^{n'} \times 2^{m'}$ can be mapped onto the hypercube of dimension $d = n' + m'$

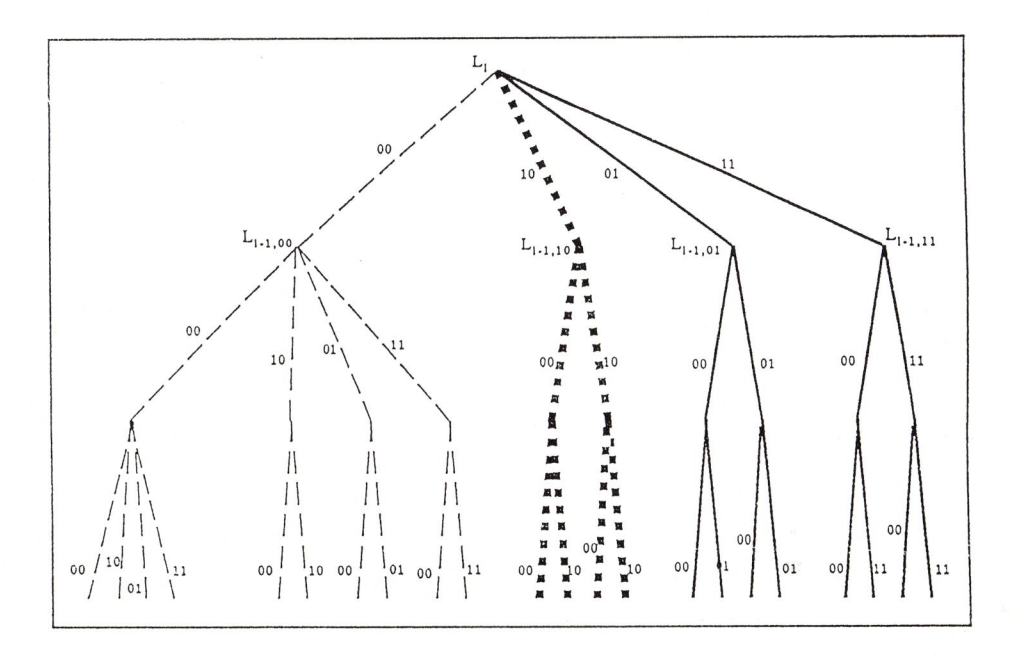

Figure 4: Necessary coarse grid tree. The dashed and dotted subtrees are computed in parallel.

- All the subconfigurations arising in the recursive subdivision process are again hypercubes of appropriate dimension and therefore have the necessary torus connections

- Using a special mapping of the (sub-) configurations onto the hypercube, it can be shown that the maximum distance for transporting parts of the defect and correction grid functions is less than or equal to one. Therefore the cost of the transport steps can be neglected, even for larger configurations.

4.3 Speedup Results

The speedups have been determined for 2, 4, 8 and 16 processors (this is due to the hypercube configuration used) and various grid sizes. The program solves Eqn. (1) with periodic boundary conditions in both directions. Therefore a grid on level l has $2^{l+1} \times 2^{l+1}$ unknowns. Table 2 shows the results for the V-cycle, the mixed cycle described above, and for the W-cycle. The speedups obtained for the mixed cycle, where the load balance is at its best, are very good even for coarse grids. However it should be noted that numerical tests have shown that the W-cycle is necessary to achieve multigrid behaviour, i.e. a convergence rate independent of h_l.

Table 2: Speedups for different cycle forms

cycle	l	2 processors		4 processors		8 processors		16 processors	
		S(2)	E(2)	S(4)	E(4)	S(8)	E(8)	S(16)	E(16)
V	3	1.82	0.91	3.26	0.81	5.43	0.68		
	4	1.90	0.95	3.58	0.89	6.28	0.79	11.00	0.69
	5			3.59	0.90	6.80	0.85	12.48	0.78
	6							13.95	0.87
mixed	3	1.93	0.97	3.59	0.90	5.82	0.73		
	4	1.98	0.99	3.80	0.95	7.04	0.88	11.89	0.74
	5			3.92	0.98	7.63	0.95	14.39	0.90
	6							15.47	0.97
W	3	1.86	0.93	3.21	0.80	5.24	0.66		
	4	1.88	0.94	3.50	0.88	5.78	0.72	9.47	0.59
	5			3.56	0.89	6.49	0.81	11.04	0.69
	6							12.29	0.77

Table 3: Comparison of Normalized Parallel Computation Times

α	β	GS − rb			FDM			ILU		
		ρ	$T_{np}(8)$	$T_{np}(16)$	ρ	$T_{np}(8)$	$T_{np}(16)$	ρ	$T_{np}(8)$	$T_{np}(16)$
1	1	0.108	0.9	0.5	0.086	5.9	3.4	0.121	0.86	0.73
$\frac{1}{2}$	2	0.393	2.23	1.1	0.196	8.8	5.2	0.150	0.96	0.81
$\frac{1}{10}$	10	0.938	32.5	16.2	0.295	11.8	6.9	0.135	0.91	0.77
10^{-2}	10^2	0.977	89.4	44.7	0.050	4.8	2.8	$8 \cdot 10^{-4}$	0.26	0.22
10^{-5}	10^5	0.977	89.4	44.7	0.051	4.8	2.8	$4 \cdot 10^{-15}$	0.05	0.05

5. Comparison and Conclusions

In order to enable an objective comparison of iterative methods on a parallel computer, we define the *normalized parallel computation time* as follows:

$$T_{np} = \frac{T_{cycle}}{-\ln \rho \cdot S(p)} \qquad (15)$$

where T_{cycle} is the time needed for one iteration on a single computer, ρ is the average convergence rate and $S(p)$ is the speedup of the method on p processors. T_{np} is the time needed for p processors to reduce the error in the solution by a factor of $1/e$.

Table 3 shows values of T_{np} obtained on 8 and 16 processors for the frequency decomposition method, and for standard multi-grid methods with ILU and Gauß-Seidel red-black smoothing applied to the anisotropic equation with $h = 1/64$ on the finest grid. Two pre-smoothing steps were used for the red-black and frequency decomposition methods, and one for the ILU method. No post-smoothing was done.

In addition V-cycles were used for the red-black and ILU measurements, and W-cycles for the frequency decomposition tests. Full-Weighting was used as restriction operator in all cases.

We make the following observations:

- the frequency decomposition and ILU methods are superior to red-black Gauß-Seidel on the parallel computer for the anisotropic equation if $\alpha/\beta > 10$ or $\alpha/\beta < 0.1$. Here the superior convergence of the two robust methods more than outweighs the better parallel efficiency of the red-black multi-grid algorithm. For example, for $\alpha/\beta = 10^{-4}$ the frequency decomposition method is 16 times faster than the standard method. ILU is 200 times as fast for the same computation on 16 processors.

- Even for Poisson's equation ($\alpha = \beta = 1$) ILU is comparable to red-black Gauß-Seidel.

- In two dimensions ILU is superior to the frequency decomposition method. This is also true for other problems tested, see [1] for numerical test results.

- ILU is not robust for the three-dimensional anisotropic equation, but the frequency decomposition method is expected to be.

- the frequency decomposition method as described in [4] cannot solve the convection dominated convection-diffusion equation, owing to the Galerkin coarse-grid equations. ILU shows in this case a similar superiority over the red-black method, see [6].

- the frequency decomposition method may be attractive for parallel computers, where each node is a vector processor, as all components are vectorizable, whereas ILU does not permit vectorization of the line segments that occur in Alg. 1.

References

[1] BASTIAN, P.: Die Frequenzzerlegungsmethode als robustes Mehrgitterverfahren: Implementierung und Parallelisierung. Diplomarbeit am IMMD III, Universität Erlangen-Nürnberg, 1989.

[2] BECKER, C., FERZIGER, J. H., PERIC, M., SCHEUERER, G.: Finite Volume Multi-Grid Solution of the Two-dimensional Incompressible Navier-Stokes Equations. in Robust Multi-Grid Methods, Proceedings of the Fourth GAMM Seminar, Notes on Numerical Fluid Mechanics, Volume 23, Vieweg Verlag, Braunschweig, 1988.

[3] HACKBUSCH, W.: Multi-Grid Methods and Applications. Springer, Berlin, Heidelberg 1985.

[4] HACKBUSCH, W.: The Frequency Decomposition Multi-Grid Method. in Robust Multi-Grid Methods, Proceedings of the Fourth GAMM Seminar, Notes on Numerical Fluid Mechanics, Volume 23, Vieweg Verlag, Braunschweig, 1988.

[5] HÄNDLER, W., MAEHLE, E., WIRL, K.: The DIRMU Testbed for High-Performance Multiprocessor Configurations. Proceedings of the first International Conference on Supercomputing Systems, St. Petersburg, 1985.

[6] HORTON, G.: Parallelisierung eines Mehrgitterverfahrens mit ILU - Glättung. Diplomarbeit am IMMD III, Universität Erlangen-Nürnberg, 1989.

[7] WITTUM, G.: On the Robustness of ILU Smoothing. in Robust Multi-Grid Methods, Proceedings of the Fourth GAMM Seminar, Notes on Numerical Fluid Mechanics, Volume 23, Vieweg Verlag, Braunschweig, 1988.

[8] WITTUM, G.: Multi-Grid Methods for Stokes- and Navier-Stokes-Equations. This volume.

[9] STONE, H. L.: Iterative Solution of Implicit Approximations of Multidimensional Partial Differential Equations, SIAM J. Numer. Anal., Vol. 5, No. 3, September 1968

THE INFLUENCE OF REENTRANT CORNERS IN THE NUMERICAL APPROXIMATION OF VISCOUS FLOW PROBLEMS

H. Blum
Institut für Angewandte Mathematik
Universität Heidelberg, D-6900 Heidelberg

SUMMARY

The effect of reentrant corners is studied for finite element discretizations of viscous flow problems. In the case of low Reynolds numbers, the *pollution effect* of the corner singularities is localized to a relatively small neighborhood of the irregular points, depending on the geometry of the domain. However, for problems with strong convection one has to expect some downstream pollution. On the basis of an asymptotic expansion of the discretization error, the computational accuracy can be significantly improved by means of extrapolation or related techniques.

1. INTRODUCTION

The accuracy of standard finite element schemes for elliptic problems on polygonal or polyhedral domains is significantly reduced due to the influence of the corner singularities. The reduction is usually not only concentrated in a small neighborhood of the irregular boundary points but extends also to interior subdomains where the solution is smooth. This so-called *pollution effect* is well studied and documented in the existing literature for several elliptic model problems including, e.g., the Poisson equation and the Kirchhoff plate bending problem.

In the present note, we deal with the influence of the corner singularities in the approximation of viscous flow problems governed by the Stokes equations

$$-\nu \Delta u + \nabla p = f, \quad \nabla \cdot u = 0 \quad \text{in } \Omega, \tag{1}$$

or by the Navier-Stokes equations

$$-\nu \Delta u + (u \cdot \nabla) u + \nabla p = f, \quad \nabla \cdot u = 0 \quad \text{in } \Omega, \tag{2}$$

together with Dirichlet boundary conditions for the velocity. Here, u and p denote the velocity and the pressure, respectively, and ν is the viscosity.

As a typical model situation, we consider the flow over a backward facing step, i.e., a channel of height $H = 1$ and length $L \gg 1$ with a reentrant corner with interior angle $3\pi/2$. In two space dimensions, the singular functions for the Stokes problem are well-known. In the nonlinear case (2) at least the dominant singularities coincide with those for problem (1), see [4]. Compared to the standard test calculations for elliptic model problems, however, the pollution effect exhibits several different phenomena which will be briefly discussed below.

The basis for our investigation is an asymptotic expansion of the discretization error which separates the influence of the singularities from that of the smooth part of the continuous solution. For standard discretizations of the Poisson equation

$$-\Delta u = f, \quad \text{in } \Omega, \quad u = 0, \quad \text{on } \partial\Omega, \tag{3}$$

such expansion have been derived in [5]. On certain uniform meshes, there holds that

$$(u - u_h)(x) = \sum_{n=1}^{N} A_n k_n(u) h_n^{2\alpha_n} s_{-1;n}(x) + O(h^{2-\varepsilon}). \tag{4}$$

Here, N denotes the number of reentrant corners z_n, $k_n(u)$ are the so-called stress intensity factors, A_n are certain constants, and h_n is the local mesh size at z_n. The exponents α_n correspond to the r-exponents of the dominant singular function, i.e. $\alpha_n = \pi/\omega_n < 1$ for problem (3), where ω_n is the interior angle of the corner. A similar result holds true for the Stokes equation, see the discussion in Section 2, below.

It is easily seen from (4) that the functions $s_{-1;n}$ serve as the *amplification factors* for the influence of the singularities at a given point x. They are defined as those solutions of the given boundary value problem which correspond to homogeneous data and grow like $O\left(|x - z_n|^{-\alpha_n}\right)$ as $x \to z_n$, see section 2 for a formal definition. Thus, close to the corner point z_n, there unavoidably occurs an error of the reduced order $O\left(h_n^{\alpha_n}\right)$. In the interior of Ω the functions $s_{-1;n}$ are smooth and the error is of the order $O\left(h^{\min 2\alpha_n}\right)$. These asymptotic result are best possible and characterize the error behavior as $h \to 0$. For practical purposes, however, one is much more interested in quantitative results for realistic sizes of the mesh parameters.

In most of the systematic numerical tests reported in the literature only a small set of model domains is considered including the L-shaped domain, a square with a slit, or a circular wedge. In all these cases, for isotropic elliptic operators the functions $s_{-1,n}$ essentially decay like the radially symmetric function $O\left(r^{-\alpha_n}\right)$ and, thus, pollution strongly influences the error in the major part of the domain. In our present problem of a channel with length $L \gg H$ the *Saint-Venant principle* applies and the effect of the singularity decays exponentially in the distance r from the corner,

$$s_{-1;n} \sim e^{-r/H}.$$

In practice, even for relatively small values of the mesh size, say $h = 1/64$ one can simply *ignore* the influence of the pollution effect already a unit distance away from the critical point. For quantitative results on the decay behavior see the figures in [2]. On uniform meshes it is even possible to increase the accuracy of the schemes outside this neighborhood by means of Richardson extrapolation with respect to the power h^2. By simply combining the discrete solutions for two different meshes, $u'_h \equiv (4u_{h/2} - u_h)/3$, extremely accurate approximations are obtained. As a consequence, it is advisable to work with fully uniform meshes without any refinements as long as one is intersted in accurate results only at some distance from the corner. This is confirmed by numerical tests presented in Section 3, below, for the Stokes problem and also for the Navier-Stokes equations in the case of relatively low Reynolds numbers ($Re \sim 50$).

In a certain neighborhood of the corner (independent of h) the pollution terms are dominant and the reduced order of accuracy becomes visible as here the dual singular functions take large values. Following the approach of [5] one may try to improve the accuracy in this part of the domain by means of extrapolation with respect to the appropriate fractional powers of h. The efficiency of this simple modification is demonstrated in Section 3, below, for the Stokes equations. In the Navier-Stokes problem, however, the influence of the boundary layer at the corner becomes visible already for small Reynolds numbers. Our numerical tests did not even show monotone convergence at all mesh points. This means that the discrete solutions are not in the *asymptotic range* of the a priori estimates for realistic mesh sizes. Correspondingly, the results obtained by fractional extrapolation will not be reliable. In practice, however, this effect appears to be of minor importance since the layers will influence the solution much stronger than the corner singularities. In such a situation one should always work with refined meshes which, then, will simultaneously reduce the corner effects.

The exponential decay of singularities described above is a typically elliptic phenomenon. For problems involving strong convection the character of the impact of the corners on the global discretization error changes significantly. In Section 4, below, we present some preliminary results for the linear convection diffusion problem

$$-\varepsilon \Delta u + \beta \cdot \nabla u = f, \quad \text{in } \Omega, \quad \varepsilon \ll |\beta|. \tag{5}$$

This equation may serve as a model for Navier-Stokes flow in case of high Reynolds numbers. As one might expect, the upstream effect of the corner singularities nearly vanishes for this type of problem. In the downstream direction one observes a very thin layer of large pollution in the direction of the tranport, while it is fully negligible outside this region. The more complicated situation of the full Navier-Stokes problem on the case of higher Reynolds numbers will be the subject of future research.

2. THE POLLUTION EFFECT FOR THE STOKES EQUATIONS

We will now describe in more detail the influence of the corner singularities in the approximation of the Stokes problem. For discretizing problem (1) we consider a standard finite element scheme based on the weak formulation with respect to the primitive variables $\{u, p\}$. For simplicity, let Ω be a polygonal domain.

Let $T_h \equiv \{K\}$ denote a regular family of subdivisions of Ω into triangular or quadrilateral elements K of diameter $O(h)$. Accordingly, let V_h and L_h be finite dimensional subspaces of $(H_0^1(\Omega))^2$ and $L_0^2(\Omega)$, respectively, consisting of piecewise polynomial functions with respect to T_h. Here, $H_0^1(\Omega)$ is the usual Sobolev space of functions possessing (generalized) derivatives in $L^2(\Omega)$, and L_0^2 consists of L^2-functions with zero mean value over Ω. The boundary values for u are assumed to be given in terms of some function $b \in (H_0^1(\Omega))^2$. Let $b_h \in V_h$ be a suitable approximation of b. Then, the finite element discretization of (1) seeks for a pair $\{u_h, p_h\} \in (b_h + V_h) \times L_h$ satisfying

$$\nu\left(\nabla(u - u_h), \nabla \varphi_h\right) - (p - p_h, \nabla \cdot \varphi_h) + (\chi_h, \nabla \cdot (u - u_h)) = 0, \tag{6}$$

for all test functions $\{\varphi_h, \chi_h\} \in V_h \times L_h$.

For a sufficiently smooth solution $\{u, p\}$, various L^2-error estimates of optimal order are known for this scheme provided that the pair of spaces $V_h \times L_h$ satisfies the uniform Brezzi-Babuška stability estimate, see [6]. Moreover, under certain additional assumptions which are usually satisfied in practice, also optimal order pointwise estimates are known, see [8].

In our numerical tests discussed below we have chosen

$$V_h \equiv \{v_h \in (H_0^1(\Omega))^2 : v_{h|K} \text{ (isoparametric) bilinear}, \forall K \in T_h, v_h \equiv 0 \text{ on } \partial\Omega\},$$
$$L_h \equiv \{r_h \in L_0^2(\Omega) : r_{h|K} \in P_0', \forall K \in T_{2h}\}.$$

By P_0' we denote the span of the piecewise constant functions on a 2 × 2-element patch which are orthogonal to the so-called local *checkerboard mode*. In several tests for cases of smooth solutions $\{u, p\}$, this Stokes element has shown very good approximation properties results, also for the pressure, see [1].

In order to understand the influence of the corner points we have to recall several results on the asymptotic behavior of the solution of the continuous problem. This problem has been studied, e.g., in Kondrat'ev [9] for a single elliptic equation and later on for general elliptic systems by Maz'ja-Plamenevskij [11]. For notational convenience, let us consider only the case of a single corner point. The solution $\{u, p\}$ of problem (1) admits the singular decomposition

$$\{u, p\} = \sum_{i=1}^{I} k_i \{s_i^u, s_i^p\} + \{U, P\}, \tag{7}$$

with a smooth remainder $\{U, P\}$. The constants k_i depend only on the data of the problem, see below. The superscripts u and p refer to the velocity and pressure components, respectively. The *singular functions* $\{s_i^u, s_i^p\}$ are particular solutions of the homogeneous Stokes problem of the form

$$\{s_i^u, s_i^p\} = \{r^{\alpha_i} t_i^u(\varphi), r^{\alpha_i - 1} t_i^p(\varphi)\}, \tag{8}$$

satisfying homogeneous Dirichlet boundary conditions along the (straight) edges close to the corner. This latter property can be achieved if and only if the values α_i are roots of the equation $\alpha^2 \sin^2 \omega = \sin^2 \alpha\omega$, where ω is the interior angle of the corner. For example, in the case $\omega = 3\pi/2$, the lowest positive solution are $\alpha_1 \sim 0.544\ldots$ and $\alpha_2 \sim 0.908\ldots$.

For our purposes, it is crucial to have precise information about the dependence of the coefficients k_i in (7) on the data f, b, and a possible inhomogeneity g in the continuity equation. Following the ideas of [11], we introduce the singular functions with negative exponents $-\alpha_i$,

$$\{\tilde{s}_{-i}^u, \tilde{s}_{-i}^p\} = c_i \{r^{-\alpha_i} t_i^u(\varphi), r^{-\alpha_i - 1} t_i^p(\varphi)\}.$$

where c_i are certain normalization constants. Clearly, these functions do not have bounded energy and, thus, cannot appear in the expansion (7). To derive a representation formula for k_i, we need related functions that exhibit the same singular behavior at the corner but satisfy homogeneous boundary conditions on all of $\partial\Omega$. To this end, we set

$$\{s_{-i}^u, s_{-i}^p\} \equiv \{\tilde{s}_{-i}^u, \tilde{s}_{-i}^p\} - \{v_{-i}^u, v_{-i}^p\}, \tag{9}$$

where $\{v_{-i}^u, v_{-i}^p\}$ are defined as the *weak* solutions of the homogeneous equations corresponding to the boundary conditions

$$v_{-i}^u = \tilde{s}_{-i}^u \quad \text{on } \partial\Omega.$$

Since the restriction of $\tilde{s}_{-i}^u$ to the boundary is a smooth function, these weak solutions are smooth in the interior of Ω. Then, from the Green identity and the expansion (7), one easily obtains the representation formula

$$k_i = (f, s_{-i}^u)_\Omega + (g, s_{-i}^p)_\Omega - (b, \partial_n s_{-i}^u)_{\partial\Omega} - (b \cdot n, s_{-i}^p)_{\partial\Omega}. \tag{10}$$

Analogous representations hold true for general elliptic systems. It turns out that the test functions in formulas of the type (10) must be homogeneous solutions of the *adjoint* boundary value problem. They are therefore called the adjoint or *dual* singular functions.

As a particular choice of the data we set $b \equiv 0$ and replace f or g in (10) by the Dirac functional. In this way, we get the following result on the singular behavior of the Green tensor with source point at x,

$$\begin{aligned} \{g_u^u(x), g_u^p(x)\} &= \sum_{i=1}^{I} s_{-i}^u(x) \{s_i^u, s_i^p\} + \{G_u^u(x), G_u^p(x)\} \\ \{g_p^u(x), g_p^p(x)\} &= \sum_{i=1}^{I} s_{-i}^p(x) \{s_i^u, s_i^p\} + \{G_p^u(x), G_p^p(x)\} . \end{aligned} \tag{11}$$

The lower subscript indicates the Green function for the velocity or the pressure, respectively. The remainder terms denoted by G contain the characteristic singularity at the source point x, but they are smooth at the corner points.

Let us now come back to the analysis of the discretization scheme (6). The following argument closely follows the discussion in [5], for the Poisson equation.

First, by linearity, we can split the error according to (7),

$$\{u - u_h, p - p_h\} = \sum_{i=1}^{I} k_i \{s_i^u - s_{i,h}^u, s_i^p - s_{i,h}^p\} + \{U - U_h, P - P_h\}.$$

If sufficiently many singular functions are split from the solution to guarantee that $U \in C^2$, i.e. if I is chosen large enough, the remainder can be estimated to be of optimal order,

$$\|U - U_h\|_\infty + h \|P - P_h\|_\infty = O(h^{2-\varepsilon}).$$

To derive this result, we have to apply the weighted norm estimates from [7] in combinations with interior pointwise error estimates as presented in [12] for scalar second order problems. On piecewise uniform meshes one even has second order superconvergence for the pressure at the centers of gravity of each element K. To treat the pointwise error for a single singular function we represent the point values of $\{s_i^u - s_{i,h}^u, s_i^p - s_{i,h}^p\}$ by using the Green tensor as a test function. Dropping the reference to the source point in the notation, we obtain that

$$s_i^u - s_{i,h}^u = \left(\nabla(s_i^u - s_{i,h}^u), \nabla g_u^u\right) - \left(\nabla \cdot (s_i^u - s_{i,h}^u), g_u^p\right) - (s_i^u - i_h s_i^u, \partial_n g_u^u)_{\partial\Omega} ,$$

where i_h denotes the usual interpolation operator in V_h. In view of the orthogonality relations (6), we have that

$$\begin{aligned} s_i^u - s_{i,h}^u &= \left(\nabla(s_i^u - s_{i,h}^u), \nabla(g_u^u - g_{u,h}^u)\right) - \left(s_i^p - s_{i,h}^p, \nabla \cdot (g_u^u - g_{u,h}^u)\right) \\ &\quad - \left(g_u^p - g_{u,h}^p, \nabla \cdot (s_i^u - s_{i,h}^u)\right) - (s_i^u - i_h s_i^u, \partial_n g_u^u)_{\partial\Omega} . \end{aligned}$$

Then, we replace the components of the Green tensor by the singular expansion (11) and arrive at

$$\begin{aligned} s_i^u - s_{i,h}^u &= \sum_{j=1}^{I} s_{-j}^u \Big[\left(\nabla(s_i^u - s_{i,h}^u), \nabla(s_j^u - s_{j,h}^u)\right) - \left(s_i^p - s_{i,h}^p, \nabla \cdot (s_j^u - s_{j,h}^u)\right) \\ &\quad - \left(s_j^p - s_{j,h}^p, \nabla \cdot (s_i^u - s_{i,h}^u)\right)\Big] + O(h^{2-\varepsilon}). \end{aligned} \tag{12}$$

This representation of the discretization error holds true on arbitrary *regular* meshes and contains the full information about the pollution effect. The corresponding pollution terms for the pressure are of course proportional to the pressure component s^p_{-j}. The several inner products in the sum on the right can be estimated to be of the order $O(h^{\alpha_i+\alpha_j})$. Thus, the term for $i = j = 1$ represents the dominant error contribution.

From (12) one can extract a more precise information if the meshes are kept uniform in a fixed neighborhood of the corner. Then, the error representation can be converted into an asymptotic expansion with respect to fractional powers of the mesh size,

$$\left(\nabla(s^u_i - s^u_{i,h}), \nabla(s^u_j - s^u_{j,h})\right) = A_{ij}h^{\alpha_i+\alpha_j} + O(h^{2-\varepsilon}),$$
$$\left(s^p_i - s^p_{i,h}, \nabla \cdot (s^u_j - s^u_{j,h})\right) = B_{ij}h^{\alpha_i+\alpha_j} + O(h^{2-\varepsilon}).$$

The main tool for the proof of this result is the scaling argument from [5] in combination with an analogue of the interior pointwise error estimates of [12]. Although a full theoretical justification of the latter estimates for the Stokes problem is not yet available in the literature, the techniques of [12] will certainly carry over to our more general situation. Finally, we remark that the inner products for $i \neq j$ are typically of higher order than indicated by the asymptotic estimate, above, and, therefore, only the pollution terms with $i = j$ will contribute to the error expansion. This can rigorously be proven in the case that the local mesh reflects the symmetry properties of the angular parts of the singular functions.

3. NUMERICAL RESULTS

In order to illustrate the results of the preceding discussion, particularly those concerning the decay properties of the pollution terms, we present some numerical tests for the approximation of a flow over a backward facing step. Further, the efficiency of Richardson extrapolation is demonstrated, at least in the "far field". Our model configuration is indicated in the following figure.

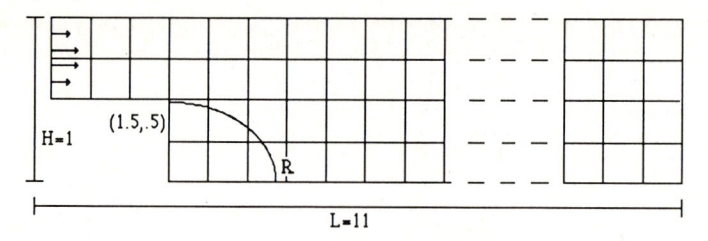

Fig. 1 Model configuration

The two problems under consideration are the Stokes equations and the Navier-Stokes equations at the low Reynolds number $Re = 50$ ($\nu = 1/100$). All calculations were performed on globally uniform meshes with mesh size h. The results, below, are partly reported from [2].

First, we consider the error in the "far field". As already mentioned before, the dual singular functions for the Stokes problem decay exponentially in the distance r from the corner. Tables 1 and 2 show typical results for the velocity and the pressure at a distance $r \geq 1$, i.e. for $x \geq 2.5$. The pressure at the vertices is defined by taking appropriate mean values of the values at the centers of gravity such that one can expect $O(h^2)$-superconvergence on the uniform mesh.

The convergence rates seem to be $O(h^2)$ in the linear as well as in the nonlinear case. This means that the discretization error for the remainder dominates the pollution terms in this part of the domain. One even can improve the results by Richardson extrapolation with respect to the power h^2. The corresponding discrete values are denoted by u'_h and p'_h, respectively.

Table 1 Stokes problem, far field results

1/h	$u_h(P)$	$u'_h(P)$	$p_h(P)$	$p'_h(P)$
8	.492065	—	9.0693	—
16	.498045	.500038	9.2734	9.3414
32	.499517	.500008	9.3271	9.3450
64	.499884	.500006	9.3423	9.3474

Table 2 Navier-Stokes problem ($\nu = 1/100$), far field results

1/h	$u_h(P)$	$u'_h(P)$	$p_h(P)$	$p'_h(P)$
8	.55958	—	8.4713	—
16	.59058	.60082	9.3559	9.6508
32	.59753	.59985	9.6010	9.6827

As a further example, we have calculated the position of the so-called reattachement point R of the zero streamline in the case $Re = 50$, see Fig. 1. Since this point is situated in the domain of weak pollution, the accuracy may be increased by h^2-extrapolation. Table 3 shows the (normalized) length of the recirculation zone, i.e. the distance of R from the step, see also [10], for the case $L = 22$.

Table 3 Navier-Stokes problem ($\nu = 1/100$), recirculation length

1/h	L_h	L'_h
8	2.795	—
16	2.279	2.107
32	2.176	2.141

Next, we study the error in the subdomain of "strong pollution" the shape of which is close to a circle around the corner of diameter 1, see [2]. Here, the values of the dual singular functions are significantly different from zero and the pollution terms of the orders $O(h^{2\alpha_i})$ dominate the error in the smooth remainder. In our case, there holds $2\alpha_1 \sim 1.08$ and $2\alpha_2 \sim 1.91$. Table 4 shows results obtained after one or two steps of extrapolation with respect to these fractional powers of h. We note that even for the Stokes problem the discrete values (for the pressure) for $h = 1/8$ seem not to be in the asymptotic range. For finer mesh sizes, however, fractional extrapolation gives us a significant improvement. The columns "ratio" contain the quotient of the errors compared to an "exact" value of the velocity at the chosen point P which was obtained from a stream function calculation on an extremely fine mesh. They show the predicted orders $O(h^{2\alpha_1})$ and $O(h^{2\alpha_2})$.

In the nonlinear case, the discrete values close to the corner are not very reliable, for the velocity as well as for the pressure. Our example in Table 5 shows that there exist vertices P of the triangulation in which one does not even observe monotone convergence for mesh sizes $h \geq 1/32$. This indicates that here the asymptotic expansion is virtually meaningless for realistic values of h. This oscillatory convergence behavior is apparently caused by the dominant influence of the boundary layer induced by the corner point. The width of the layer in our example is about $1/50$, i.e. nearly as small as our finest mesh size. Therefore, if in the nonlinear case one is really interested in discrete values close to the corner it is advised to use locally refined meshes or certain damping strategies. This would then simultaneously also reduce the effect of the elliptic corner singularity.

Table 4 Stokes problem, polluted domain

$1/h$	$u_h(P)$	ratio	$u'_h(P)$	ratio	$u''_h(P)$	$p_h(P)$	$p'_h(P)$	$p''_h(P)$
8	.708886	—	—	"exact"$\sim$.68219		21.515	—	—
16	.690853	.324	.674839	—	—	21.523	—	—
32	.685376	.368	.680512	.229	.682763	21.402	21.295	—
64	.683425	.387	.681692	.298	.682161	21.360	21.323	21.334

Table 5 Navier-Stokes problem ($\nu = 1/100$), polluted domain

$1/h$	u_h	p_h
8	1.03090	6.2922
12	.99938	6.4807
16	.99225	6.4224
24	.98547	6.4402
32	.99677	6.4882

4. THE POLLUTION EFFECT IN CONVECTION-DIFFUSION PROBLEMS

In this final section we present some preliminary considerations concerning the character of the pollution effect in the presence of strong convective terms. We consider the usual scalar model problem

$$-\varepsilon \Delta u^\varepsilon + \beta \cdot \nabla u = f \quad \text{in } \Omega, \qquad u^\varepsilon = b \quad \text{on } \partial\Omega, \tag{13}$$

where, for simplicity, we assume that $\beta \in \mathbf{R}^2$ is constant. For moderate values of ε we choose the standard Galerkin approximation, defining $u_h^\varepsilon \in b_h + V_h$ by the relation

$$\varepsilon(\nabla u_h^\varepsilon, \nabla \varphi_h) + (\beta \cdot \nabla u_h^\varepsilon, \varphi_h) = (f, \varphi_h) \quad \forall \varphi \in V_h. \tag{14}$$

In our test calculations for small ε, we have used streamline upwinding for the convective term to avoid the typical oscillatory behavior of central differences.

The coefficients k_i for this problem could in principle be obtained from the representation formulas for the Poisson equation applied to the right hand side $\varepsilon^{-1}(f - \beta \cdot \nabla u)$. In order to avoid the explicit dependence on the unknown solution u we have to build in the low order derivative into the definition of the dual singular functions. To this end, we set

$$s_{-i}^{\varepsilon,-\beta} \equiv \varepsilon^{-1}(\tilde{s}_{-i} - v_{-i}). \tag{15}$$

where $\tilde{s}_i = (i\pi)^{-1} r^{-\alpha_i} \sin \alpha_i \varphi$ is a (suitably normalized) singular solution of the homogeneous Poisson equation. Here, the correction v_{-i} is defined as the weak solution of the adjoint boundary value problem (with reversed direction of β)

$$\varepsilon(\nabla v_{-i}, \nabla \varphi) + (-\beta \cdot \nabla v_{-i}, \varphi) = (-\beta \cdot \nabla \tilde{s}_{-i}, \varphi) \quad \forall \varphi \in H_0^1(\Omega), \tag{16}$$

where $v_{-i} \equiv \tilde{s}_{-i}$ on the boundary. Then, using again the Green identity, we can derive the following representation of the singular coefficients,

$$k_i = \varepsilon^{-1} \left(f, s_{-i}^{\varepsilon, -\beta} \right) - \left(b, \partial_n s_{-i}^{\varepsilon, -\beta} \right)_{\partial \Omega}. \tag{17}$$

The corresponding formula for the adjoint operator is obtained by replacing $-\beta$ by $+\beta$ in the definition of v_{-i}. Then, choosing $b \equiv 0$ and $f \equiv \delta$ as before, we see that the singular coefficients of the Green function for problem (13) are just of the form $\varepsilon^{-1} s_{-i}^{\varepsilon, \beta}$. By an argument analogous to that sketched in Section 2 we can then extract from the error the dominant pollution term,

$$(u^\varepsilon - u_h^\varepsilon)(x) \sim k_1(u^\varepsilon) \|\nabla (s_1 - s_{1,h})\|^2 s_{-1}^{\varepsilon, \beta}(x). \tag{18}$$

For large viscosity, $\varepsilon \sim |\beta|$, the size of k_1 and the distribution of the dual singular function are comparable to those for the Poisson equation. For small ε, however, the values of $\tilde{s}_{-1}$ and v_{-1} nearly coincide outside possible layers of width $O(\varepsilon)$. These layers are caused by incompatible data at the outflow boundary or by the influence of the right hand side. In our case, the singular data at the corner lead to large values of $s_{-1}^{\varepsilon, \beta}$ at distance $O(\varepsilon)$ which then are transported in downstream direction without significant damping.

To illustrate these effects we have calculated approximations for the dual singular functions on an L-shaped domain for $\beta = (1,1)$ for various values of ε. The example given in the following figure shows the case $\varepsilon = 10^{-4}$, where the significant values are concentrated in a strip of size h in transport direction.

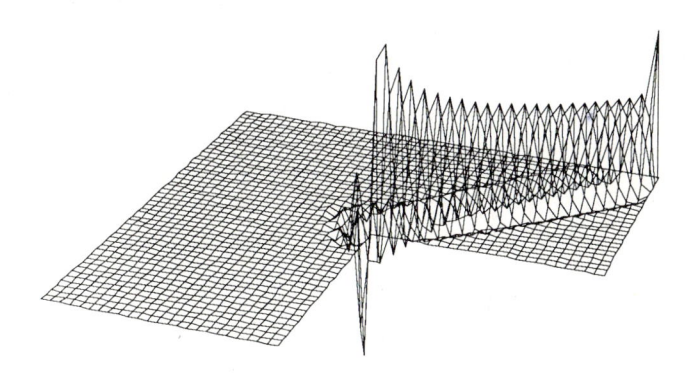

Fig. 2 Dual singular functions for $\varepsilon = 10^{-4}$

Correspondingly, from (17) we see that the value of k_1 is essentially determined by the data f and b in *upstream* direction. Moreover, the maxima of the dual singular functions outside a neighborhood of the corner slightly increase with decreasing ε which also affects the size of the singular coefficient for fixed data.

The foregoing observations lead us to the following conclusions about the pollution effect in convection dominated problems.
(i) The corner influence can be visible only in a narrow strip downstream in the characteristic direction. In this subdomain the error may even become larger for small ε while outside pollution can be fully *neglected*. The effect can be removed by suitable modifications of the discretization schemes, e.g., by fractional extrapolation, or by local mesh refinements. In view of (18), an optimally designed mesh has to reduce the energy error of the first singular function, $\|\nabla(s_1 - s_{1,h})\|$. Therefore, the same refinement strategy should be used as for the the Poisson problem, i.e. refinement is needed only at the corner and *not* in the transport direction.
(ii) Due to the non-local character of the pollution terms such an optimal mesh will in general not be generated by standard self-adaptive strategies. Most of these techniques are based on local error estimation and, therefore, will lead to refinements in the whole downstream area.

This latter problem and the related effects in the case of non-constant or even nonlinear convective terms will be studied elsewhere.

REFERENCES

[1] Blum, H., Harig. J., Keller, G., Müller, S.: *A numerical comparison of approximation and stability properties of low order Stokes elements*, Preprint SFB 123, Univ. Heidelberg 1989, to appear

[2] Blum, H., Harig. J., Müller, S.: *Extrapolation techniques for finite element approximations of Stokes and Navier-Stokes flow over a backward facing step*, in: Finite Approximations in Fluid Mechanics II (Hirschel, E., ed.), Notes on Numerical Fluid Mechanics **25**, Vieweg 1989

[3] Blum, H., Lin, Q., Rannacher, R.: *Asymptotic error expansion and Richardson extrapolation for linear finite elements*, Numer. Math. **49**, 11–37 (1986)

[4] Blum, H., Rannacher, R.: *On the boundary value problem of the biharmonic operator on domains with angular corners*, Math. Meth. Appl. Sci. **2**, 556–581 (1980)

[5] Blum, H., Rannacher, R.: *Extrapolation techniques for reducing the pollution effect of reentrant corners in the finite element method*, Numer. Math. **52**, 539–564 (1988)

[6] Brezzi, F.: *On the existence, uniqueness, and approximation of saddle-point problems arising from Lagrange multipliers*, RAIRO, Anal. Numer. **R2**, 129–151 (1974)

[7] Dobrowolski, M.: *Numerical approximation of elliptic interface and corner problems*, Habilitationsschrift, Univ. Bonn (1981)

[8] Duran, R., Nocchetto, R.H., Wang., J.: *Sharp maximum norm error estimates for finite element approximations of the Stokes problem in 2-D*, SIAM J. Numer. Anal., to appear

[9] Kondrat'ev, V.A., *Boundary value problems for elliptic equations in domains with conical or angular points*, Trans. Moscow Math. Soc. **16**, 227–313 (1967)

[10] Morgan,K., Periaux, J., Thomasset, F. (eds.) *Analysis of laminar flow over a backward facing step*, Notes on Numerical Fluid Mechanics **9**, Vieweg 1984

[11] Maz'ja, V.G., Plamenevskij, P.A., *Coefficients in the asymptotics of the solutions of elliptic boundary value problems*, J. Sov. Math. **9**, 750–764 (1980)

[12] Schatz, A.H., Wahlbin, L.B.: *Interior maximum norm estimates for finite element methods*, Math. Comput. **33**, 465–492 (1979)

A FINITE VOLUME DISCRETIZATION WITH IMPROVED ACCURACY FOR THE COMPRESSIBLE NAVIER-STOKES EQUATIONS[1]

Mart Borsboom
Delft Hydraulics
P.O.Box 152, 8300AD Emmeloord, The Netherlands

An optimal low-order conservative scheme on a three-by-three point computational molecule has been developed, for discretizing the two-dimensional steady-state compressible Navier-Stokes equations. The scheme is almost free of numerical diffusion, even on highly distorted grids, while still relatively easy to program. It is therefore considered to be a useful compromise for Navier-Stokes calculations.

INTRODUCTION

A numerical solution method for the Navier-Stokes equations should be such that it is capable of generating a good approximation to the exact solution of this differential problem (provided that a solution exists under the specified conditions), at a minimal cost. This means that an easy-to-program, low-order discretization scheme, that is generally applicable and rather insensitive to the grid structure, is to be prefered.

Especially the last point is of crucial importance [1]. The number of unknowns is proportional to the number of grid cells, and should be kept as small as possible in order to minimize the computer time needed to solve the discretized system of equations. On the other hand, the grid should also provide a sufficiently high resolution. If we consider for a moment the complicated structures that may appear in compressible, viscous flows, we get a good appreciation of what that signifies for the grid design.

In many fields of application, small regions of steep gradients may be present in the flow domain, even when the variables have been Reynolds-averaged. To capture such details numerically, local concentrations of grid points are necessary, while at the same time the total number of points should be minimized. This can only be achieved by using highly distorted grids, but then a low-order scheme may not be sufficiently accurate anymore [1-5]. This makes the generation of grids that are suited to numerical viscous flow simulations a very difficult task, and in practice one always tries to keep the grid as smooth as possible (see e.g. [1,2]). It will be clear that the optimal way to handle this problem will be some sort of compromise, and we cite Thompson, who says that "at the same time that effort is made to generate better grids, a similar effort should be made to develop hosted algorithms that are more tolerant of the grids" [1].

---

[1] This work has been carried out at the Von Karman Institute for Fluid Dynamics, Rhode-St-Genese, Belgium.

In this paper we propose to discuss to some extend the numerical modeling of the steady-state Navier-Stokes equations on an arbitrary grid. An analysis technique will be presented that was found to be very useful in determining the "optimal" compromise.

LOW-ORDER CONSERVATIVE SCHEMES ON A NON-UNIFORM GRID

It is a straightforward matter nowadays, to design discretization methods for use on a uniform grid, that are both accurate and simple. In practice, however, the usefulness of uniform grids is rather limited, especially for Navier-Stokes calculations. We will therefore analyse in this section the use of some one-dimensional schemes on a non-uniform grid, and extend the analysis to discretization methods for two-dimensional differential equations (i.e. the Navier-Stokes equations) on non-uniform grids in the next section.

Some loss in accuracy may be expected, and our objective is to limit this loss as much as possible, while still keeping the discretization scheme relatively simple. For this reason, only low-order discretization schemes will be investigated, that discretize the equations on a computational molecule of (three by) three grid points. For a one-dimensional non-uniform grid, such a molecule is given in Figure 1, together with the notations that will be employed. A *low-order scheme* will be defined as a discretization scheme that is obtained using an at most piecewise linear representation of the unknowns, whereby the order of the equations is normally reduced by writing the equations in an integral form and applying Gauss' theorem.

We will restrict our investigations to centered discretization schemes. Non-centered schemes always introduce some form of artificial dissipation in the numerical model of a convection-diffusion equation, that may obscure the physical dissipation effects completely [6,7]. However, it seems that at present it is still not possible to construct and to use computational grids that are fine enough to avoid the necessity of the introduction of any artificial dissipation, when high Reynolds number flows are to be simulated numerically. In other words, "a suitable finite difference scheme for a high Reynolds number flow will be one that maintains good accuracy in the convection-dominated region; in the regions where the convection terms and the viscous terms in the original differential equation should balance each other, which cannot be done exactly numerically, the truncation error should enhance the weight of the viscous terms." (Shyy [8]). When the truncation error is enhanced by means of some added artificial dissipation (as is always the case when *central* discretizations are used), one has at least some idea about the magnitude of the modeling error that, unfortunately, had to be made. We consider the insight that may be gained from such a "smoothing" procedure to be of great value [6,7].

Consider now the linear differential equation $d\Phi/dx = f$, written in the integral form:

$$\int_V d\Phi = \int_V f \, dx , \qquad (1)$$

which equation should hold for any finite volume V belonging to the domain of definition of the unknown function Φ and the source term f.

Discretized over the finite volumes $[x_{j-1}, x_{j+1}]$, equation (1) yields:

$$\Phi^h_{j+1} - \Phi^h_{j-1} = \int_{x_{j-1}}^{x_{j+1}} f \, dx , \qquad (2)$$

with Φ^h_j the numerical approximation Φ^h of Φ at grid point x_j.

Clearly, the discretization is exact if the integral of f will be evaluated exactly. However, for numerical purposes, also the right-hand side of (2) should be discretized, and the simplest way to do this is by approximating f over the whole interval $[x_{j-1}, x_{j+1}]$ by its value at the point x_j. This yields the discretization:

$$\Phi^h_{j+1} - \Phi^h_{j-1} = (h_j + h_{j+1}) f_j , \qquad (3)$$

which would also have been obtained by transforming the physical coordinate x to a computational coordinate ξ such that $x = x(\xi)$, and by discretizing the differential equation $d\Phi/dx = f$, transformed to the computational ξ-domain, by means of finite differences, whereby the transformation coefficient $dx/d\xi$ is approximated by the centered difference $(x_{j+1} - x_{j-1})/2\Delta\xi$. The accuracy of discretization (3) for the *transformed* equation is $O(\Delta^2\xi)$, since centered discretizations have been used on a grid that is uniform in ξ, where $\Delta\xi$ denotes the constant length of each grid cell.

The order of accuracy of (3) on the *original* domain with variable grid spacing in x can be obtained by determining the order of magnitude of the *truncation error* of the discretization. For that purpose Taylor series expansions can be used, and in order to find out around which point(s), it suffices to realize that the integral in (2) can be approximated equally well by means of a Gauss' integration formula. For example, if a piecewise constant or a piecewise linear approximation of the integrand is used in order to arrive at a discretized form of (2) that will be at most third-order accurate, the integral may as well be approximated by means of a one-point quadrature rule, that is also third-order accurate. This yields:

$$\int_{x_{j-1}}^{x_{j+1}} f \, dx = 2\bar{h} f(\bar{x}) + O(\bar{h}^3) , \qquad (4)$$

and we conclude that the finite volume discretization (3) is equivalent to a finite difference discretization times $2\bar{h}$ of $d\Phi/dx = f$ at the point $x = \bar{x} = (x_{j+1} + x_{j-1})/2$, that is of the same order of accuracy. A Taylor series expansion around this point shows that (3) approximates the left-hand side of (1) up to at least $O(\bar{h}^3)$, while the approximation of the right-hand side yields an error term of $O(\bar{h}h_j - \bar{h}h_{j+1})$. Using the derivative of the equation itself in order to eliminate the first derivative of f, the equivalent differential equation of (3) is obtained:

$$\frac{d\Phi^h}{dx} = f + \frac{h_j - h_{j+1}}{2} \frac{d^2\Phi^h}{dx^2} + O(\bar{h}^2) . \qquad (5)$$

It follows that the truncation error of (3) is $O(\Delta\bar{h}, \bar{h}^2)$, where $\Delta\bar{h}$ denotes the difference in grid spacing over one computational molecule.

At first sight, the conclusion would be that discretization (3) is first-order accurate on a non-uniform grid, and second-order accurate on a uniform grid. However, the former conclusion is only valid if we suppose that $\Delta\bar{h}$ is proportional to the grid size $\bar{h}$, for any value of $\bar{h}$. This may be true for some strange construction of the grid, but in practical applications the grid will normally be constructed in such a way that any grid refinement will cause at the same time a grid uniformization. So it will be more realistic to say that $\Delta\bar{h}$ will be at least proportional to $\bar{h}^2$, and that (3) will be a second-order accurate discretization of (1).

This corresponds also to the order of accuracy of (3) for the transformed problem in the computational ξ-domain. It can easily be verified that if the uniform grid in ξ is refined, the non-uniform grid in the x-domain becomes indeed more uniform!

Notice however that, although in practice scheme (3) will almost always behave as a *second-order* accurate scheme, the fact remains that its truncation error is only *first order* in $\Delta\bar{h}$. This implies that, because of dimensional reasons, the coefficient of the second-derivative error term will *only* be zero when discretization (3) is used on a uniform grid, but that it may be very large on a non-uniform grid, indicating the presence of *numerical diffusion*. Although this may have not too drastic consequences in case the equation $d\Phi/dx = f$ is to be solved numerically, it may deteriorate completely the accuracy of the numerical approximation when the discretization of a second derivative is added to the scheme in order to model numerically some convection-diffusion phenomenon.

As mentioned before, it may be necessary to add artificial dissipation (e.g. second-derivative error terms) to a numerical scheme for reasons of *accuracy*. Obviously, these terms should be kept as small as possible, if we want to study numerically equations containing physical dissipation terms. So any source that contributes to this error in an uncontrollable way should be reduced to a level at which its influence will be negligible. Numerical diffusion is such a source, and should therefore be eliminated as much as possible, especially for Navier-Stokes calculations. It is a quantity that in general is very difficult to control, because it depends on the structure of the grid and may change sign. It would therefore be best to use discretization schemes *and* grids that are introducing as less numerical diffusion as possible.

For scheme (3) this would imply that the use of an (almost) uniform grid is imperative. But this has the consequence that we lose all the freedom in grid design, which is definitely not desirable. Alternatively, we may also try to develop a discretization scheme for (2), that does *not* introduce that much numerical diffusion, when applied on a non-uniform grid. This can be achieved by using more accurate interpolations for the integrand f. The resulting discretization scheme is:

$$\Phi^h_{j+1} - \Phi^h_{j-1} = h_j \frac{f_{j-1}+f_j}{2} + h_{j+1} \frac{f_j+f_{j+1}}{2}, \tag{6}$$

in case of a piecewise linear approximation of f, and:

$$\Phi^h_{j+1} - \Phi^h_{j-1} = \qquad (7)$$

$$= (h_j + h_{j+1}) \left[\frac{2h_j - h_{j+1}}{6h_j} f_{j-1} + \frac{(h_j + h_{j+1})^2}{6 h_j h_{j+1}} f_j + \frac{2h_{j+1} - h_j}{6 h_{j+1}} f_{j+1} \right],$$

if a quadratic approximation is used. The latter scheme is actually a higher-order scheme, but has been included here for illustration purposes.

Because discretization (6) has been obtained using second-order accurate, linear interpolations, its truncation error can still be found by means of a Taylor series expansion around the single quadrature point $x = \bar{x}$. This shows that the approximation error of the scheme is $O(\bar{h}^2)$, *independently* of the grid structure, and does not contain any numerical diffusion.

With regard to scheme (7), it may be assumed that this discretization of (2) will be the most accurate one possible on a three-point computational molecule. Since it is based on a third-order accurate interpolation of f, and since it is fourth-order accurate on a uniform grid, we expect its truncation error to be $O(\bar{h}^2 \Delta \bar{h}, \bar{h}^4)$.

Because of the higher-order interpolation of f, a higher-order quadrature rule is to be used in order to find the relation between this global finite volume discretization and pointwise finite difference discretizations:

$$\int_{x_{j-1}}^{x_{j+1}} f \, dx = \bar{h} f(\bar{x} - \bar{h}/\sqrt{3}) + \bar{h} f(\bar{x} + \bar{h}/\sqrt{3}) + O(\bar{h}^5) \ . \qquad (8)$$

This shows that it should be possible in principle to split (7) in *two* finite difference discretizations at the *two* quadrature points $x = \bar{x} \pm \bar{h}/\sqrt{3}$, such that a Taylor series expansion of each of those will yield the expected truncation error. This appeared indeed to be possible, which proves that in the truncation error of (7) neither second-derivative error terms, nor third-derivative error terms appear, while the fourth-derivative error term vanishes on a uniform grid.

A rather remarkable discretization scheme is obtained by adding one third of scheme (3) to two thirds of scheme (6). It can easily be verified that this scheme is fourth-order accurate (and identical to scheme (7)) on a *uniform* grid, but nevertheless *not* free of numerical diffusion when used on a non-uniform grid, because of scheme (3).

The discretization we talk about is a *finite element* scheme, obtained by means of the low-order Galerkin finite element discretization (GFEM) of $d\Phi/dx = f$, using piecewise linear basis functions. This proves that, although the finite element technique yields very accurate discretization schemes on *uniform* grids, the technique may be completely unsuited for use with non-uniform grids (see also [6,9]). A similar conclusion applies also to the finite difference method, and to certain finite volume schemes. For this reason such techniques should preferably be used on smoothly varying grids when convection-dominated flow problems (e.g. the Navier-Stokes equations at high Reynolds number) are to be solved [1-5].

The analytical results of this section have all been verified by means of a series of numerical tests, discretizing and solving approximately the linear equation (1) on a non-uniform grid with a $\Delta \bar{h}$ of order $\bar{h}$. It was found that the theory predicts and explains the observed error behavior very well, *including* the errors due to the discretization of the numerical boundary condition [10].

THE OPTIMAL LOW-ORDER CONSERVATIVE SCHEME FOR THE STEADY-STATE NAVIER-STOKES EQUATIONS

Discretization methods in space for the steady-state compressible Navier-Stokes equations in two dimensions will be considered. The system of equations we will be dealing with may be written in the general conservative form:

$$\frac{\partial \underline{F}}{\partial x} + \frac{\partial \underline{G}}{\partial y} = \frac{\partial}{\partial x}\left[V_1\frac{\partial \underline{V}_2}{\partial x} + V_3\frac{\partial \underline{V}_4}{\partial y}\right] + \frac{\partial}{\partial y}\left[W_1\frac{\partial \underline{W}_2}{\partial x} + W_3\frac{\partial \underline{W}_4}{\partial y}\right]. \qquad (9)$$

The vectors $\underline{F}$, $\underline{G}$, $\underline{V}_i$ and $\underline{W}_i$, and the matrices V_i and W_i are in general functions of the vector of unknowns $\underline{U}$ and the space coordinates x and y. The precise form of these vectors and matrices is irrelevant for the development of a discretization method, and is also rather difficult to give, as (9) not only comprises the continuity, momentum, and energy equation, but may also contain differential equations concerning the turbulence model, chemical reactions, two-phase flow, etcetera.

The integration of (9) over some finite volume V belonging to its domain of definition yields:

$$\oint_S \underline{F} dy - \oint_S \underline{G} dx = \oint_S \left[V_1\frac{\partial \underline{V}_2}{\partial x} + V_3\frac{\partial \underline{V}_4}{\partial y}\right] dy - \oint_S \left[W_1\frac{\partial \underline{W}_2}{\partial x} + W_3\frac{\partial \underline{W}_4}{\partial y}\right] dx, \qquad (10)$$

where the Gauss' theorem has been applied in order to reduce the order of the integrands, and where S is the contour of V.

The object of this section is to put equation (10) in a discretized form in such a way that an *accurate* numerical approximation of the solution of (10) can be obtained by solving the resulting system of algebraic equations. This is a rather difficult thing to ask, since the accuracy of a discretization is not only determined by the discretization method that has been employed, but depends also on the grid point resolution *and* on the smoothness of the grid. Both grid aspects are important, but here we will only treat the latter one, by trying to develop a discretization scheme that is sufficiently accurate for Navier-Stokes calculations, independently of the grid structure.

This means in the first place that the amount of spurious diffusion introduced by the numerical scheme should be sufficiently small, which is a rather severe demand if viscous flows at high Reynolds number are to be simulated numerically. On the other hand we would also like to use a simple low-order scheme, which may inevitably lead to the introduction of some numerical diffusion. The question is: does an acceptable compromise exist?

In general, a low-order discretization method may only be expected to be exact (and hence free of any truncation error) for functions that are linear in the coordinates x and y, indicating that the truncation error will normally contain error terms of second and higher derivatives [6,9]. So for a low-order discretization of (10), the equivalent equation is expected to be of the form:

$$\frac{\partial F^h}{\partial x} + \frac{\partial G^h}{\partial y} = \frac{\partial}{\partial x}\left[(1+c_{21}^{21})V_1^h\frac{\partial V_{-2}^h}{\partial x} + c_{22}^{21}V_1^h\frac{\partial V_{-2}^h}{\partial y}\right] + \frac{\partial}{\partial y}\left[c_{23}^{21}V_1^h\frac{\partial V_{-2}^h}{\partial x} + c_{24}^{21}V_1^h\frac{\partial V_{-2}^h}{\partial y}\right] +$$

$$+ \frac{\partial}{\partial x}\left[c_{21}^{22}V_3^h\frac{\partial V_{-4}^h}{\partial x} + (1+c_{22}^{22})V_3^h\frac{\partial V_{-4}^h}{\partial y}\right] + \frac{\partial}{\partial y}\left[c_{23}^{22}V_3^h\frac{\partial V_{-4}^h}{\partial x} + c_{24}^{22}V_3^h\frac{\partial V_{-4}^h}{\partial y}\right] +$$

$$+ \frac{\partial}{\partial x}\left[c_{21}^{23}W_1^h\frac{\partial W_{-2}^h}{\partial x} + c_{22}^{23}W_1^h\frac{\partial W_{-2}^h}{\partial y}\right] + \frac{\partial}{\partial y}\left[(1+c_{23}^{23})W_1^h\frac{\partial W_{-2}^h}{\partial x} + c_{24}^{23}W_1^h\frac{\partial W_{-2}^h}{\partial y}\right] +$$

$$+ \frac{\partial}{\partial x}\left[c_{21}^{24}W_3^h\frac{\partial W_{-4}^h}{\partial x} + c_{22}^{24}W_3^h\frac{\partial W_{-4}^h}{\partial y}\right] + \frac{\partial}{\partial y}\left[c_{23}^{24}W_3^h\frac{\partial W_{-4}^h}{\partial x} + (1+c_{24}^{24})W_3^h\frac{\partial W_{-4}^h}{\partial y}\right] +$$

$$+ c_{21}^{11}\frac{\partial^2 F^h}{\partial x^2} + c_{22}^{11}\frac{\partial^2 F^h}{\partial x \partial y} + c_{23}^{11}\frac{\partial^2 F^h}{\partial y^2} +$$

$$+ c_{21}^{12}\frac{\partial^2 G^h}{\partial x^2} + c_{22}^{12}\frac{\partial^2 G^h}{\partial x \partial y} + c_{23}^{12}\frac{\partial^2 G^h}{\partial y^2} + \text{HOT} .$$

(11)

The numerical approximations of the functions in (9) are denoted by the superscript h. The super- and subscripts of the c_{jl}^{ik}-coefficients indicate that it concerns the l-th j-th-order derivative truncation error term, of the discretization of the k-th i-th-order derivative physical term in (9).

The c_{jl}^{ik}-coefficients are only functions of the coordinates x and y, because the difference operators should, like the differential operators, be linear in the unknown functions. Actually, this applies only when F^h, G^h, etc. are sufficiently smooth. In order to cope with steep gradient regions in the solution, it may be necessary to make the scheme solution-dependent, i.e. nonlinear, in order to obtain the required local enhancement of the truncation error. We prefer however to refer to these enhancements (caused by the use of flux limiters or smoothing terms, for example), as *artificial* truncation errors, introduced in order to *improve* the accuracy of the numerical solution. This is in contrast with the *numerical* truncation errors due to the linear discretization scheme, that usually *limit* the accuracy of the solution.

The expressions for the c_{jl}^{ik} can be obtained by means of Taylor series expansions around quadrature points (see previous Section, and [9,11]). This technique can therefore be used to derive in an analytical way a low-order conservative discretization scheme, that introduces no or only a minimal amount of numerical diffusion in the equivalent equation (11).

Such an analysis is in fact only practical for low-order schemes, where it suffices to expand the whole discretization formula around one single quadrature point.

To begin with, we will consider the general discrete approximations of the convective flux integrals of (10):

$$\oint_S \underline{F}\, dy \approx \sum_i \alpha_i \underline{F}_i^h \,, \quad \oint_S \underline{G}\, dx \approx \sum_i \beta_i \underline{G}_i^h \,, \tag{12}$$

where the discretizations are based on the values $\underline{F}_i^h$ and $\underline{G}_i^h$ of the approximation functions $\underline{F}^h$ and $\underline{G}^h$ at a certain number of grid points (x_i, y_i). Since the differential operators $\partial/\partial x$ and $\partial/\partial y$ are linear, the α_i and the β_i should all be functions of the coordinates x_i and y_i only.

We will not fix any finite volume shape, and will define the point (x^*, y^*) to be equal to the (unknown) quadrature point. A Taylor series expansion of (12) around (x^*, y^*) yields the equivalent differential forms of the discretizations up to $O(h^2 Vol)$, with h the size of the grid and Vol the surface of the finite volume, and therefore the general expressions for the c_{21}^{1k} that are $O(\Delta h)$. For the discrete approximation (12) to be free of numerical diffusion, it is necessary that these coefficients all vanish, but also the truncation error terms of lower order should vanish.

This procedure leads to a number of conditions to be satisfied by the α_i and the β_i [6,11]. It follows that the α_i and β_i should be of the form:

$$\alpha_i = \sum_j \theta^{ij} y_j \,, \quad \beta_i = \sum_j \theta^{ij} x_j \,, \quad \sum_j \theta^{ij} = 0 \,, \tag{13}$$

where the θ^{ij} are constants of $O(1)$ to be determined. It can now be proven that if we approximate the convective flux terms of every finite volume of the computational domain by the same discretized form (12), by keeping the θ^{ij} in (13) constant over the whole domain (i.e. by using a low-order discretization technique based on piecewise linear interpolations), it will in general *not* be possible to have a diffusion-free approximation of $\partial \underline{F}/\partial x$ and $\partial \underline{G}/\partial y$ on every finite volume. In other words, on an arbitrary grid *any* low-order conservative discretization of first derivatives causes the presence of spurious diffusion terms in the equivalent differential equation [6,11]. For such a discretization scheme, the coefficients c_{21}^{1k} in (11) are indeed in general not equal to zero, and may therefore pose a serious problem in *any* numerical solution method for the Navier-Stokes equations at high Reynolds number (see also [12]).

If we would put certain restrictions to the choice of the points (x_i, y_i), the conditions that the α_i and β_i have to satisfy may be fulfilled, and the discretization may then be completely free of any numerical diffusion. Such restrictions may however not be possible in practice, since the distribution of points in a computational grid also has to suit the physical problem to be modeled. On a uniform grid, for example, any second-order accurate, low-order discretization of (9) of practical interest is free of numerical diffusion, but the utility of such a regular grid for Navier-Stokes calculations is of course about negligible. Alternatively, it would also be possible to make the θ^{ij}-values variable, and to use a finite volume discretization technique where these coefficients are functions of the grid points (x_i, y_i), i.e. of the shape of the computational molecule. The only way to realize this would be by using a higher-order conservative discretization scheme, but this would make a computer code considerably more complex, and would demand storage space and a calculation procedure for all the θ^{ij}.

For these reasons the possibility of molecule-dependent θ^{ij} to obtain diffusion-free conservative discretizations for convective terms on *any* two-dimensional grid has not been considered any further. Instead, we have limited our investigations to the development of an optimal low-order conservative scheme on a *structured* grid, and looked for the most accurate $\partial F/\partial x$- and $\partial G/\partial y$-discretization possible with fixed θ^{ij}, on a computational molecule of three by three points. The latter restriction is in order to keep not only the numerical scheme, but also its implementation (relatively) simple, which is especially of importance if in the future the extension to three-dimensional problems will be made.

The choice of a nine-point molecule means that, starting from the general form (12)-(13), 72 coefficients θ^{ij} are to be determined. By imposing that the discretization scheme should be both conservative and centered, quite a few degrees of freedom can be eliminated, and we are left with only 15 coefficients. Additionally, the scheme should at least be free of consistency errors, i.e. only second-derivative truncation error terms as in (11) may possibly appear. This leads to some more relations that the coefficients have to satisfy, obtained from the Taylor series expansion of (12) around the (still unknown) quadrature point. This results in a class of *six* linearly independent finite volume discretizations; any linear combination of these six schemes will also be free of consistency errors, but not of spurious numerical diffusion, due to the restrictions that we have put on this group (constant θ^{ij}-coefficients).

Fortunately, this does not have to be a serious drawback for practical applications. The design of a computational grid will often be such that highly irregular and arbitrary-looking shapes will not be present, or this situation can at least always be avoided when a structured grid is used. We may assume that in that case grid lines will be in general (nearly) parallel, at least locally, or will deviate only slightly and incidentally from that "ideal" situation.

When the grid is composed of grid cells of parallelogram shape, only *two* schemes do not introduce numerical diffusion. The other schemes do, as well as any linear combination of them. So obviously, the most accurate low-order conservative discretization scheme on a three-by-three-point computational molecule is to be sought within the group of linear combinations of these two schemes. By means of an order-of-magnitude estimation of the errors, introduced in both schemes by means of a small off-parallel deformation of the molecule, the *optimal combination* could be determined. Not surprisingly, it corresponds to a finite volume scheme with the fluxes approximated by means of isoparametric, quadratic interpolations, which seems to be about the most accurate conservative discretization possible on a structured nine-point molecule. We will refer to this optimal combination as *OLOCS*: the Optimal Low-Order Conservative Scheme. Its description can be found in [6,9,11].

The numerical diffusion coefficients of OLOCS are compared with those of GFEM in [6,9,11], for the grid that we have used to solve the steady viscous flow over a backward facing step. Apart from the step corner where both methods are of about equal (in)accuracy and where the grid could have been designed better, OLOCS introduces orders of magnitude less numerical diffusion than the equivalent GFEM, based on the same computational molecule.

Concerning a discretization method for the diffusion terms of (9), it is not possible to derive in the same rigorous way as for the convective terms an optimal scheme, due to the nonlinear character of the optimization problem. Instead, we have followed a more heuristic approach, by trying to combine optimal accuracy with ease of programming. Again, two discretization schemes were found that do not introduce numerical diffusion (a consistency error!), when the grid cells have parallel sides. The optimal combination of the two schemes, that can be found in [11], typically introduced an at most 1 % viscosity error in our calculations of the viscous flow over a backward facing step.

LAMINAR VISCOUS FLOW CALCULATIONS

The above described low-order discretization scheme for the steady-state Navier-Stokes equations has been used for several numerical flow simulations. To this end, the method was completed with a boundary procedure based on characteristic theory, and with an optimized implicit time-marching solution algorithm [11,13]. No artificial dissipation of any kind had to be added for these low-Reynolds number applications.

One of the first test cases was the compressible, low-subsonic, steady laminar flow over a backward facing step, for the four different combinations of geometry and Reynolds number that were proposed by GAMM for the workshop on numerical simulation of incompressible flow [14]. A comparison with the results of other workers shows clearly the advantages of using the developed discretization scheme. Because a highly distorted grid can be used, a high concentration of grid points around the step corner could be realized, while still keeping the total amount of points rather small (a grid of 60 times 25 points was used). This permitted to calculate numerically the very steep pressure gradients at the separation corner [6,9,11], the existence of which has been proven analytically [15], but (almost) never predicted numerically [14]. We also predicted correctly the flow separation slightly after the step corner.

The method has also been applied for the calculation of viscous flow through a cascade of NACA0012 profiles at a Reynolds number of 200, based on mean inlet velocity and chord length. Overall results can be found in [7,11], but here we would like to discuss some interesting physical details that we were able to predict numerically with the present method.

A close-up at the leading edge of the grid of 78 times 39 points that was used for these calculations is shown in Figure 2. The non-dimensionalized chord length and pitch of the blades was equal to 1, with a stagger angle of $30°$, while the positions of the inlet and the outlet plane were at -1.55 and 3.55 respectively. Reservoir conditions as well as inlet flow angle were specified at the inlet, while the static pressure was imposed at the outlet. No-slip conditions were applied at the solid wall boundaries, that were assumed to be adiabatic.

The calculated non-dimensionalized total pressure field at an inlet flow angle of $30°$ and a non-dimensionalized outlet static pressure of 0.95 is shown in Figure 3. For this case, the flow is almost symmetric with respect to the profile, with the stagnation point close to the leading edge. The calculated average Mach number at the inlet turned out to be 0.2144, while the total mass flow error was only 0.0068 %! The latter result shows clearly the excellent conservation properties of the scheme.

The decrease in total pressure that occurs as soon as the boundary layer starts to develop, is a result that was to be expected. The predicted increase at the stagnation point was however quite surprising. It can be explained physically, since due to the strong curvature of the flow around the stagnation point, a shear force exist in cross-flow direction, causing the transport of momentum from the main stream toward the stagnation point. So it is a purely viscous effect, that would not (or at least should not!) occur in an inviscid flow calculation.

The calculated total pressure increase can also be explained in a more quantitative way, by considering the analytical approximation of a two-dimensional viscous stagnation-point flow. The following approximate relation can be derived for the total pressure at the stagnation point (p_{STAG}) and at the inlet ($p_{TOT\infty}$) [16,p179]:

$$p_{STAG} = p_{TOT\infty} + \frac{B \, D \, (\rho U)_\infty}{Re} , \qquad (14)$$

with B the velocity gradient parallel to the wall at the stagnation point, D the chord length, $(\rho U)_\infty$ the average mass flux at the inlet, and Re the Reynolds number. By assuming that the flow close to the stagnation point is roughly comparable with a potential flow around a circular cylinder, the value of B can be approximated by $B = 2U_\infty/R$, with R the radius of curvature of the profile at the stagnation point.

For the presented calculation, the non-dimensionalized value of R appears to be equal to .0158, although it increases rapidly away from the stagnation point. With this value, the analytical approximation of the stagnation pressure increase appears to be .0394, using the numerically calculated values for $(\rho U)_\infty$ and U_∞. The numerically predicted increase is .0147, which is of the same order of magnitude, although smaller than expected. Most likely, this is due to the rather inaccurate estimation of R, whose effective value may be considerably higher. This explanation is confirmed by the results of the flow calculation at an inlet angle of 0°, where the stagnation point was found to be at a considerable distance from the leading edge, in a region where the radius of curvature of the blade contour varies smoothly. Here, the theory predicts a non-dimensional total pressure increase of .0037, while the numerical result shows an increase of .0041, i.e. an agreement within 10 %!

Another variable that shows an interesting behavior in the vicinity of the stagnation point is the entropy. A local decrease in entropy may be expected, due to the increase of the total pressure. The transport equation of the entropy s reads:

$$\rho T \frac{Ds}{Dt} = \frac{1}{2\mu}(\bar{\bar{\tau}} . \bar{\bar{\tau}}^T) + \underline{\nabla}(k \underline{\nabla} T) , \qquad (15)$$

where ρ, T, μ, $\bar{\bar{\tau}}$, and k denote respectively the density, the static temperature, the viscosity, the viscous stress tensor, and the thermal conductivity. It shows that the entropy may change along streamlines because of viscous dissipation and thermal conduction, the first and the second term in the right-hand side of (15) respectively.

The first term, which is positive definite, describes the irreversible rate of entropy increase. But the second term, which may be both positive *and* negative, can cause an entropy increase as well as an entropy decrease. Its effect depends on the magnitude of the coefficient *k*. Flow calculations at two different Prandtl numbers, keeping all other parameters the same, show that the numerically predicted entropy variations are indeed in qualitative agreement with (15) (Figure 3). Also the decrease of the thickness of the thermal boundary layer at larger Prandtl numbers can be observed.

CONCLUDING REMARKS

From the presented analysis it may be concluded that the relative magnitude of the truncation error rather than the order of accuracy of the applied discretization scheme determines how well a numerical flow solver will accurately predict the solution of the system of differential equations at hand. Actually, the order of accuracy of a scheme is only well defined on a uniform grid, and for sufficiently smooth solutions [6,9].

For the numerical simulation of the Navier-Stokes equations at high Reynolds number, this means in the first place that only a very small amount of numerical diffusion can be tolerated. Higher-order truncation error terms should also be sufficiently small, and indicate therefore in some sense the size of the grid that should be used. In case a sufficiently high concentration of grid points cannot be afforded, artificial dissipation should be added locally to the numerical truncation error, in order to model rather than calculate the steep gradients of the solution.

It has been proven that for a conservative discretization scheme, the introduction of numerical diffusion may only be avoided completely by using higher-order schemes. Here, we have followed a different approach, by trying to minimize rather than eliminate these low-order truncation errors. This has led to the development of an easy-to-program, low-order, conservative scheme that is almost free of numerical diffusion and fairly insensitive to the grid design, which has been demonstrated by the calculation of the flow over a backward facing step and of the flow through a cascade of NACA0012 profiles. However, some restrictions to the grid are still present, and the use of a so-called structured grid is imperative.

It is interesting to notice that on a uniform grid OLOCS is fourth-order accurate. It can be shown that OLOCS is the low-order limit of a higher-order scheme, and therefore still very accurate on a smoothly varying non-uniform grid, with a leading truncation error term of $O(\Delta^2 h)$ rather than $O(h^2)$. This nice property could be exploited fully in a multiple-block flow solver, using a structured and smooth grid per block, and a higher-order scheme at the block interfaces only. The discretization of the viscous terms will then still be at most second-order accurate, but this can hardly be considered to be a disadvantage. Firstly, because this truncation error is confined to the small layers where the viscous effects are important, and secondly because in practical applications the viscous modeling will be inaccurate anyway due to the use of turbulence models and added artificial dissipation.

ACKNOWLEDGEMENTS

The author would like to thank Prof. Herman Deconinck of the Von Karman Institute and Doctor P. Niederdrenk of the DFVLR Gottingen, for their suggestions concerning the physical interpretation and the analytical verification of the numerically obtained results.

REFERENCES

[1] THOMPSON J.F. (1984), "Grid Generation Techniques in Computational Fluid Dynamics", AIAA-J. 22, 1505-1523.
[2] THOMPSON J.F., WARSI Z.U.A., MASTIN C.W. (1985), *Numerical Grid Generation*, North-Holland, New-York.
[3] CULLEN M.J.P. (1982), "The Use of Quadratic Finite Element Methods and Irregular Grids in the Solution of Hyperbolic Problems", J.Comput.Phys. 45, 221-245.
[4] CULLEN M.J.P. (1983), "Analysis of Some Low-Order Finite Element Schemes for the Navier-Stokes Equations", J.Comput.Phys. 51, 273-290.
[5] GRESHO P.M., CHAN S.T., LEE R.L., UPSON C.D. (1984), "A Modified Finite Element Method for Solving the Time-Dependent, Incompressible Navier-Stokes Equations. Part 1: Theory", Int.J.Numer.Meth.Fluids 4, 557-598.
[6] BORSBOOM M. (1986), "Numerical Modeling of Navier-Stokes Equations", in: *Numerical Techniques for Viscous Flow Calculations in Turbomachinery Bladings*, Von Karman Institute LS1986-02.
[7] BORSBOOM M. (1987), *A Numerical Solution Method for the Steady-State Compressible Navier-Stokes Equations with Application to Channel and Blade-to-Blade Flow*, Doctoral Thesis, Université de Liège, Belgium.
[8] SHYY W. (1985), "A Study of Finite Difference Approximations to Steady-State, Convection-Dominated Flow Problems", J.Comput.Phys. 57, 415-438.
[9] BORSBOOM M. (1986), "About the (In)Accuracy of Low-Order Conservative Discretization Schemes", in: *Notes on Numerical Fluid Dynamics, Volume 13*, proceedings of the Sixth GAMM-Conference on Numerical Methods in Fluid Mechanics, Rues D., Kordulla W., eds., Vieweg, Braunschweig.
[10] BORSBOOM M. (1989), to be published.
[11] BORSBOOM M. (1988), "An Implicit Approximate Factorization Finite Volume Technique with Improved Accuracy for the Compressible Navier-Stokes Equations", in: *Computational Fluid Dynamics*, Von Karman Institute LS1988-05.
[12] ROE P.L. (1987), "Error Estimates for Cell-Vertex Solutions of the Compressible Euler Equations", ICASE Report 87-6.
[13] BORSBOOM M., STUBOS A.K., THEUNISSEN P.-H. (1988), "The Optimal Time Step for the Implicit Approximate Factorization Scheme - Theory and Applications", in: *Notes on Numerical Fluid Dynamics, Volume 20*, proceedings of the Seventh GAMM-Conference on Numerical Methods in Fluid Mechanics, Deville M., ed., Vieweg, Braunschweig.
[14] GAMM-workshop on *Analysis of Laminar Flow Over a Backward Facing Step* (1984), Morgan K., Periaux J., Thomasset F., eds., Vieweg, Braunschweig.
[15] LADEVÈZE J., PEYRET R. (1974), "Calcul Numérique d'une Solution Avec Singularité des Equations de Navier-Stokes: Ecoulement Dans un Canal Avec Variation Brusque de Section", Journal de Mécanique 13, 367-396.
[16] WHITE F.M. (1974), *Viscous Fluid Flow*, McGraw-Hill, New York.

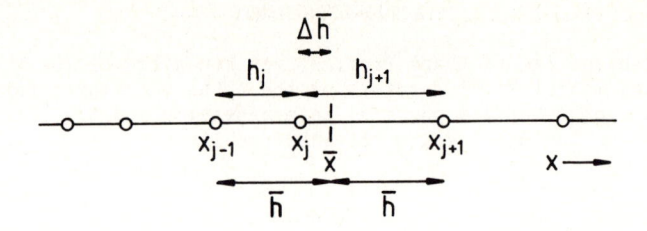

FIGURE 1 – A Three-Point Computational Molecule on a One-Dimensional, Non-Uniform Grid

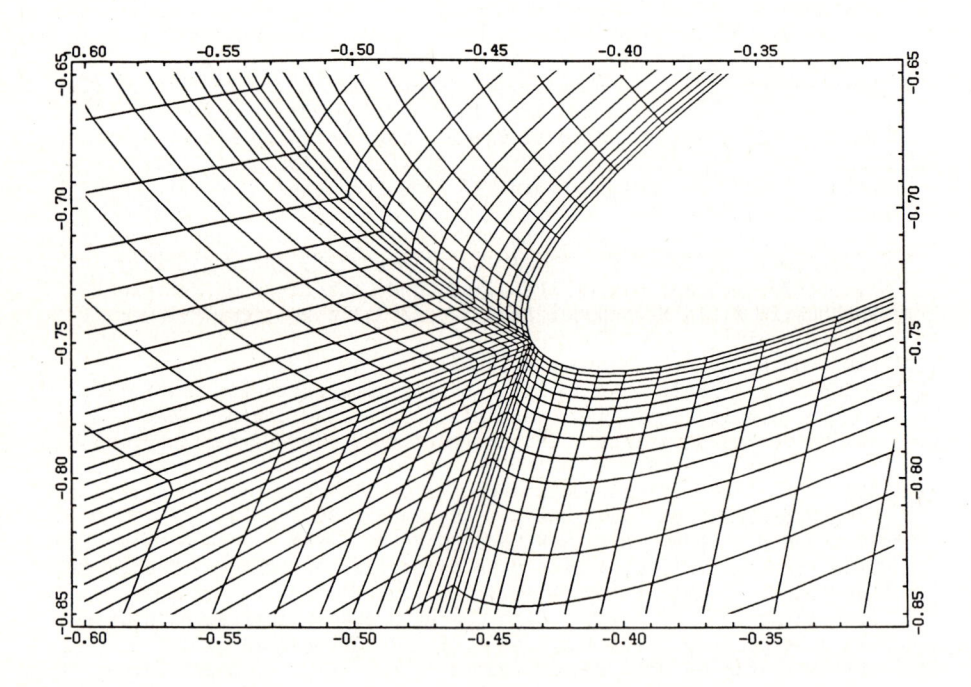

FIGURE 2 – Detail of the 78 by 39 Point Grid for the Blade-to-Blade Flow Calculations

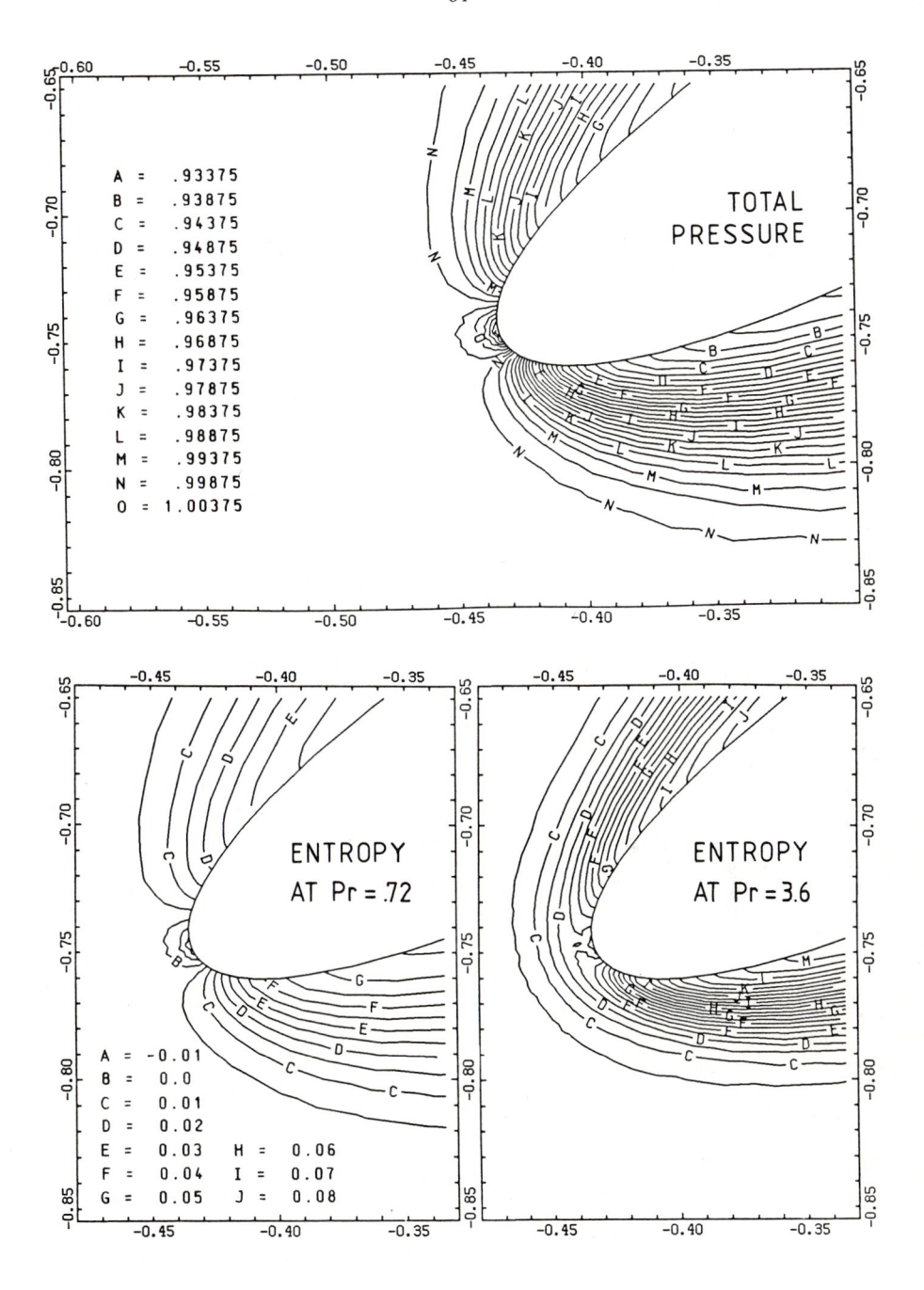

FIGURE 3 - Close-up of Viscous Steady-State Solution at Leading Edge for the Case $Re = 200$, $p_s = .95$, and $\alpha = 30°$

CALCULATION OF VISCOUS INCOMPRESSIBLE FLOWS IN TIME-DEPENDENT DOMAINS

Laszlo Fuchs

Department of Gasdynamics,
The Royal Institute of Technology,
S-100 44 Stockholm, Sweden.

and

Scientific and Technical Computing Group, ACIS
IBM Svenska AB,
S-163 92 Stockholm, Sweden.

SUMMARY

A numerical scheme for computing viscous incompressible flows on zonal grids is presented. The scheme allows the usage of local body fitted grids that overlap global cartesian grids. The computational domain may be multiply connected and contain several solid bodies. A Multi-Grid scheme, that is capable of handling zonal and locally refined grids has been developed. The convergence rate of the scheme, under different conditions, is studied. The application of the method to some flow problems exhibits the basic features of the scheme.

INTRODUCTION

The flow of a viscous incompressible liquid in a confined domain is considered. The domain may contain several solid bodies and it may be composed of several different branches. Flows in such geometries is rather common in engineering. The common numerical approach for such problems is to generate a body-fitted mesh so that solid surfaces become coordinate lines. Such typical grid generations techniques are described in [1]. An application of the method for the computation of the creeping flow past a cylinder placed in a channel is given in [2]. An alternative grid generation method is using a zonal (domain decomposition) technique. Local meshes are generated in each zone, independently of each other. The meshes in the different zones may overlap one or more meshes that belong to other zones. The problem can be solved by computing alternatingly a smaller problem in each zone and then updating the 'internal boundary' values. This scheme is not new and it is known as Schwarz algorithm (see e.g. [3,4]). An extension of the zonal algorithm is to include local mesh refinement in certain (sub-) zones. This type of local mesh refinement has been applied to viscous flows in [5] and to compressible inviscid flows in [6,7]. The application of local grids, super-imposed on a global grid, for hyperbolic equations is described

by Berger [8]. The basic approach in [7] is to use a global
body fitted mesh and to refine it locally, by halving the mesh
spacing at some sub-regions. In [8], only cartesian grids are
used. Curved shocks are embedded in several small rectangular
boxes. In each of these boxes a local cartesian mesh is used.
With all the variants of zonal grid methods one has to resolve
several difficulties:

A. Defining a proper data structure so the data can be accessed
easily and with little 'overhead'. In addition, the data
management should allow addition of zones (i.e. extending the
computational domain) and local mesh refinements. The former
requirement is of importance if the boundary conditions are not
known exactly at a given location of the computational
boundary.

B. Information exchange procedure across the zonal interfaces.
This procedure may require that certain properties (i.e.
compatibility conditions such as total mass balance) has to be
satisfied, for the existence of a solution. In addition, the
procedure must be stable, for the convergence of the numerical
algorithm.

C. The solution algorithm must be such that the convergence
rate is not affected by the usage of zonal grids. The slow
convergence rate of the basic Schwarz algorithm seems to be the
underlying reason for the limited usage of zonal-grid
techniques. One of the reasons for the renewed interest in the
zonal technique is because it lends itself naturally for
parallel computing. The aim of this paper is to demonstrate the
application of zonal-grid techniques, together with the
efficiency and flexibility of that scheme compared to other
more conventional methods even on serial computers.

Here, we consider an extended Multi-Grid method applied to
a zonal grid system. The basic grid system is composed of a
uniform cartesian mesh and several local body-fitted meshes.
The basic zonal-grid system may be extended by adding locally
refined meshes that 'cover' only parts of some of the zones
(i.e. local mesh refinements). Each of these local meshes is
derived from its 'parent' grid that contains it. The data
management is flexible enough so that new zones and new locally
refined sub-grids can always be added. The 'inter-zonal'
information exchange is described in some details and its
effect on the convergence rate is studied. The presence of
solid bodies inside the computational domain requires special
care to maintain the efficiency of the algorithm.

In the following we describe the application of the zonal
Multi-Grid (MG) algorithm for the computation of two-
dimensional viscous incompressible flows.

GOVERNING EQUATIONS AND THEIR APPROXIMATION

For 2-D flows the governing equations can be written in terms of the streamfunction and the vorticity. The two PDE's require two conditions on the whole boundary. On solid boundaries the velocity vector vanishes, providing the two required conditions. No slip boundary conditions mean that the surface of solid bodies describe a streamline and that the derivative of the streamfunction normal to the body surface vanishes. The reference value of the streamfunction is arbitrary. If, however, there are solid bodies in the flow field the (constant) value of the streamfunction on such surfaces must be determined as part of the solution. To determine this value uniquely, an additional condition must be satisfied. This condition is derived from the momentum equations and is found by using the fact that the value of the pressure at any point must be independent of the path of integration [9].

At inflow and outflow boundaries one may specify the and the vorticity themselves (instead of the velocity vector). On a partitioned (zonal) domain, one must add 'inter-zonal-boundary conditions'. Such conditions must be defined so that the problem is well posed in each zone. In our case this would imply that two types of data must be given. There is, however, some freedom in choosing such data. One may choose the components of the velocity vector or the pair streamfunction and vorticity or the combinations of these. Using the velocity vector implies a compatibility condition on each sub-grid; i.e. that the total mass flux vanishes. Since this condition is not satisfied explicitly and since it is satisfied only at convergence, this choice of inter-zonal exchange results in a substantial slow-down of the convergence of the iterative solver. Therefore, the streamfunction and the vorticity are used to exchange information across the overlapping grids.

A zonal grid system is constructed by generating a local mesh in each zone. Uniform and cartesian grids are used as much as possible with the possible exception near solid bodies, where a local (body-fitted) grid is generated. The governing equations are approximated by finite-differences on each grid, using central differences for all the terms and upwind differences for approximating the first derivatives of the vorticity. The boundary vorticity (on surfaces of solid bodies) is computed by using the no-slip condition.

DATA ORGANIZATION AND MANAGEMENT

The current zonal scheme is an extension of the local mesh refinement technique that is described in [5]. The present scheme includes the treatment of locally and independently defined (overlapping) grids and multiply-connected domains. The scheme maintains the capability of accommodating locally refined subgrids. These refinements cannot, however, be defined arbitrarily. If an arbitrarily shaped region is to be refined, a completely new zone has to be added (during the solution procedure).

The data management must be sophisticated enough to allow the flexibility that is described above. We achieve this flexibility through a system that is naturally derived from the basic MG scheme [5]: A hierarchy of grids is defined. This hierarchy consists of a set of levels and each level contains one or more sub-zones. Each sub-zone must have some region in common with at least one another sub-zone. The mesh topology in each sub-zone of the same level may be different but it is the same in each sub-zone in different levels. The mesh spacing in each sub-zone is halved when going from a given level to a finer one. It should be pointed out that not all sub-zones have to be defined in each level. Sub-zones may be multiply-connected domains, and node points that belong to unused regions are flagged by a logical variable. A typical zonal-grid (only one level) is shown in Fig. 1.a. The schematic representation of that grid system is given in Fig. 1.b. All dependent variables and auxiliary data are stored in separate vectors. Data belonging to the cartesian mesh is stored from A to B and the polar regions are stored at C-D and E-F segments. The data of the given level is stored contiguously from memory addresses A to F. The next level is stored in a similar manner at address F+1. A pointer system is defined so that each sub-grid can be accessed directly by using either the 'level index' (m) and the 'sub-zone index' (n, in Fig. 1.b) or a (uniquely defined) 'grid index' ,$k=k(m,n)$. Further pointers are required to access easily 'parent' sub-grids (for efficient MG computations). Sub-zones that employ cartesian grids require only minimal additional storage: The size, origin, angle of rotation and the number of node points in that sub-zone. For general sub-grids, we store the auxiliary data resulting from the (local) transformation of the coordinates. This data is retrieved, when needed by using an independent pointer system.

A distinction between physical and 'internal' boundaries is achieved through flagging 'internal' boundary points. Unused regions of multiply-connected zones are also marked. These markings are also used to identify the points for which data must be fetched from overlapping zones.

In all, the grid management system is more complicated than that in simpler MG solvers. Nevertheless the total overhead is small compared to the total memory needed for storing the dependent variables. This is why the overlapping grid approach is a practical method for calculating flows past complex geometries.

NUMERICAL PROCEDURE

The discrete approximations to the governing equations are solved by a straightforward MG method: Both equations are 'smoothed' by using a Successive Point Relaxation (SPR) procedure. The non-linear problem is transferred to coarser grids (using the FAS scheme, see e.g. [10]) and the coarse grid corrections are interpolated linearly to the fine grid. Details of the basic scheme are given in [11]. The extensions that are added to the basic MG scheme include the treatment of zonal grids, possibilities for local mesh refinements and a procedure for updating the streamfunction value on the surface of inner bodies.

The basic Schwarz algorithm implies the solution of the discrete equations, in each zone separately and then allowing inter-zonal exchange. The procedure is repeated until convergence is achieved. Such updating procedure results in non-smooth residuals and results in slow convergence. The scheme that we employ transfers data among the zones on the finest grid and on the coarse grids. After an inter-zonal exchange, the approximation is smoothed out by some relaxation sweeps. On the coarsest grid only the corrections are interpolated among the zones. In contrast to the current correction scheme that is used on coarse grids, one may use a FAS variant (similar to the one described below for updating the boundary value of the streamfunction). The FAS variant has not yet been implemented in our code.

As noted above, by interpolating the streamfunction and the vorticity instead of the velocity vector, no additional conditions are needed for the existence of a solution. Therefore, we may use simple interpolations for information exchange among the zones. In our code, it is possible to chose either a bi-cubic or a bi-quintic interpolation scheme. In fact, the lower order scheme is adequate on the coarse grids (where the corrections are interpolated), even when the bi-quintic interpolation is used on the finest mesh. When locally refined grids are present, the interpolation of data from the coarser grid to the boundaries of the refined mesh is simpler. In such cases, the node points on the boundaries of refined regions lie on a coordinate line in common to the next coarser grid. (This is so by the way we introduce locally refined meshes). Therefore, a one-dimensional interpolation scheme can be used on these boundaries.

The value of the streamfunction at the surfaces of different objects is computed, when it cannot be set to a given value, only on the coarsest grids. Details of the scheme are to be described elsewhere [9].

COMPUTED EXAMPLES

First, we consider the effect of some crucial steps in the scheme on the convergence rate. As computational problem we consider the flow past a single cylinder placed in a channel with an orifice at the inflow plane. The Reynolds number was 300 (based on the width of the channel). First a solution was computed, using zero values of vorticity and streamfunction as initial approximation. The well converged solution was than used to study the convergence rate using one method at a time. When boundary conditions on the inter-zonal interfaces and the value of the streamfunction on the surface of the cylinder are given, the convergence rate of the basic MG scheme is found. For simply connected domains (both cartesian and polar) the convergence factor is about 0.65 (i.e. the residual decreases by such a factor per computational effort equivalent to one sweep on the, local, finest grid). For multiply connected rectangular grids the convergence is somewhat slower (convergence factor of 0.7). The convergence history is unchanged even when the surface value of the streamfunction is updated by the procedure described above. When inter-zonal information exchange is allowed the convergence slows down on both zones (0.89). The full problem converges also with this rate. For comparison we consider the convergence of the full problem using a single grid relaxation scheme. The improved efficiency of the zonal-MG scheme is evident. The convergence histories of the different cases are given in Figure 2.a (for the rectangular mesh) and Figure 2.b (for the cylindrical mesh). The results show clearly the weakest link in the process, namely namely the inter-zonal exchange procedure. It is believed that by using the FAS variant also in this step the convergence rate would improve. This procedure has not yet been implemented, however.

We have considered several cases with two or more tubes placed in straight channel. In these calculation both steady and unsteady flows have been calculated. The unsteady cases were computed by using the basic steady-state solver, together with an implicit (trapezoidal rule) time-marching scheme. The streamlines in such a (steady flow) case are shown in Figure 1.

The flow in a square cavity, induced by a rotating disc, was studied using the above described scheme. The body-fitted grid rotates with the disc, while a cartesian grid is fixed to the walls of the cavity. The grid and the instantaneous streamilnes at six distinct time steps (the time interval between two frames equals to 1/8 of a period) are shown in Figure 3. At the first frame one can stil observe that the streamlines on the two different grids do not perfectly overlapp (not fully converged). As convergence is established, the streamlines on the different grids overlapp completely.

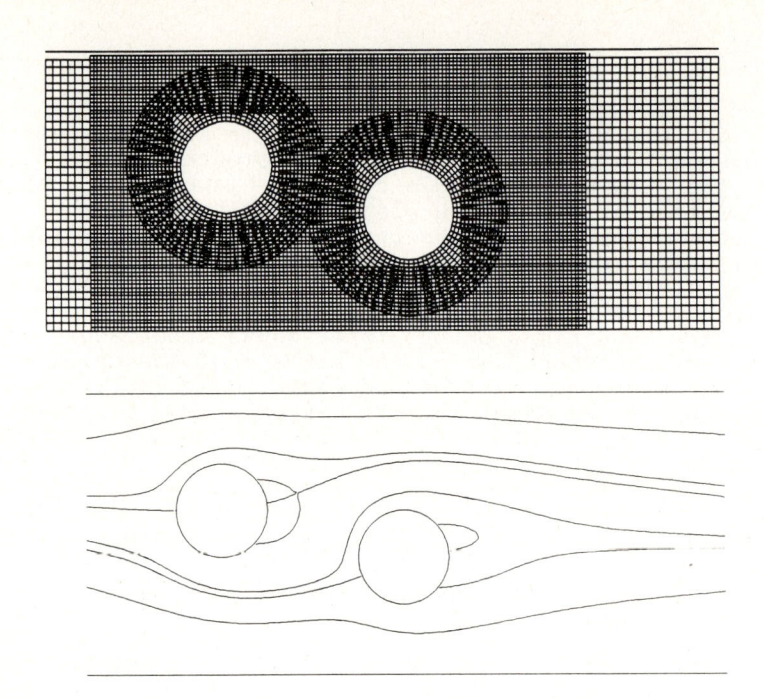

Figure 1.a: A typical zonal-grid system. The two cyliders are wrapped with cylindrical grids that overlapp the cartesian grid. The streamlines for Re=100, are plotted.

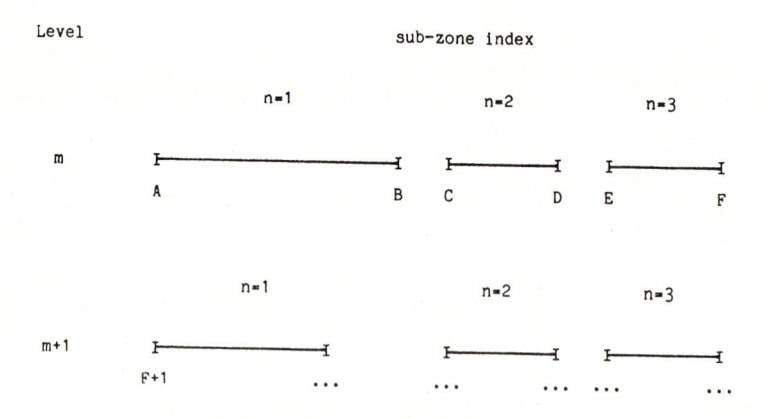

Figure 1.b: A schematic presentation of data storage for the grid in Fig. 1.a (one level). Data is stored contiguously. A finer level (m+1) may be obtained by global or local mesh refinement. The 'level index' (m) and the 'sub-zone index' (n) define the 'grid index' k.

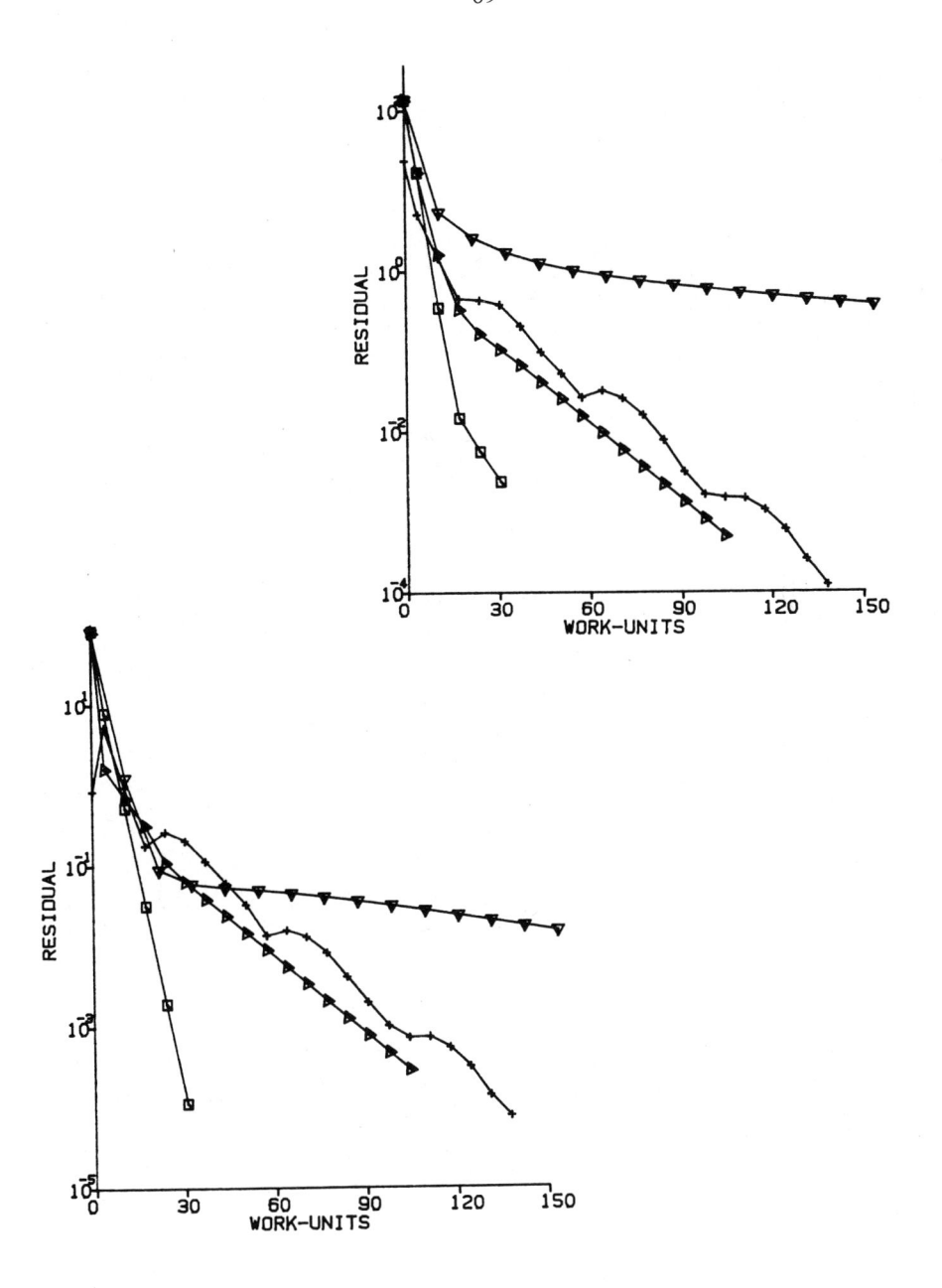

Figure 2: Convergence histories of the scheme on the cartesian grid (above) and the polar grid (below):
   ▫ - Basic Multigrid scheme (no inter-zonal update).
   ▷ - With inter-zonal update.
   + - Full Zonal-MG scheme.
   ▽ - Single grid scheme.

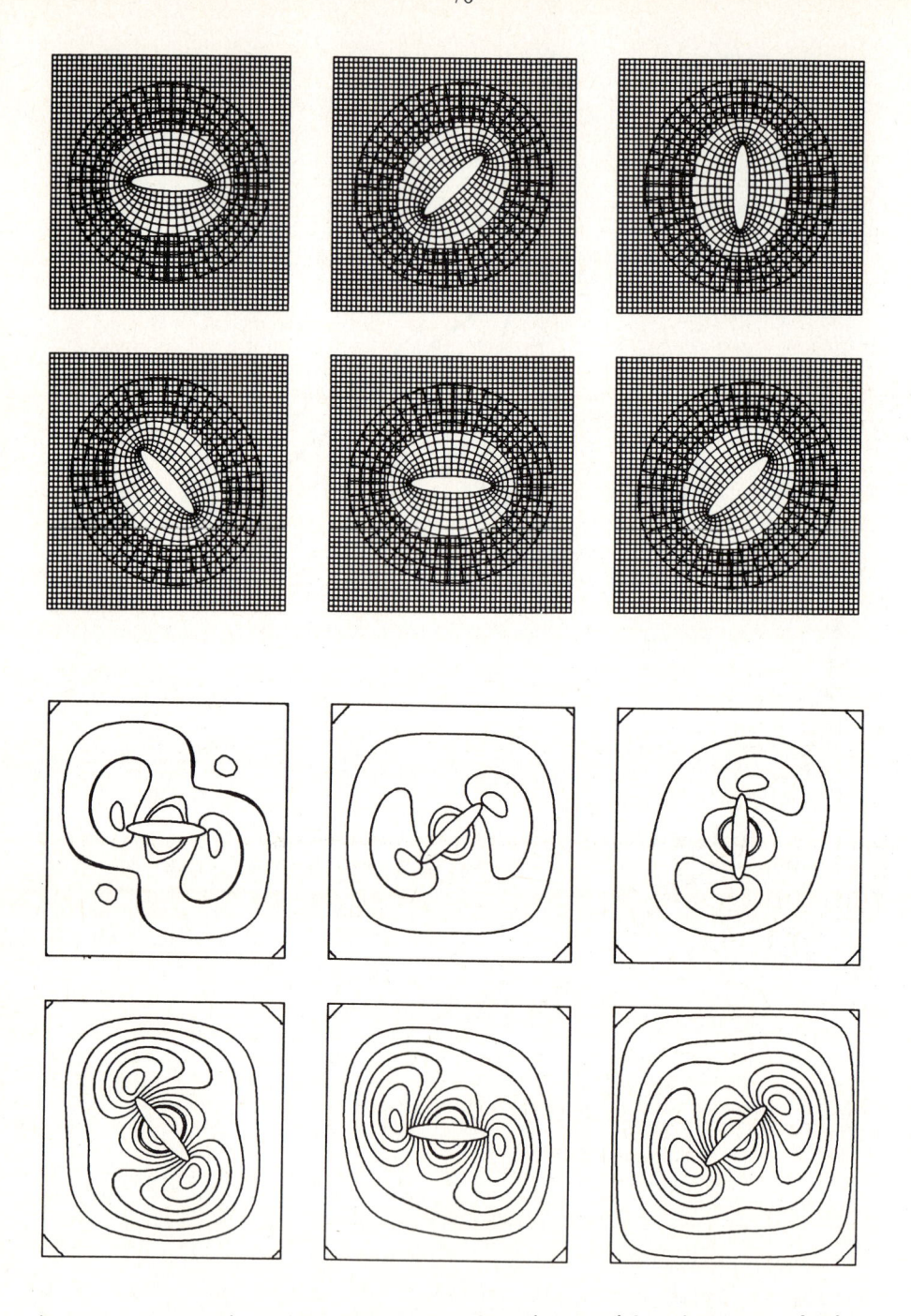

Figure 3: The time-dependent overlapping grid (above) and the instantenous streamlines. The disc rotates at a constant rate.

Concluding remarks

The main features of the zonal-MG scheme that has been developed are: Flexibility in treating complex and possibly non-fixed geometries, as well as having the possibility to adding local mesh refinements. The scheme has good numerical efficiency due to the use of local body fitted grids, the use of uniform grids (as much as possible) and the MG procedure that yields good convergence rate.

The method has been applied to some two-dimensional flows where the efficiency and the flexibility of the scheme could be studied. The major weakness of the current method is still in the inter-zonal information exchange procedure. As indicated above, the usage of a FAS scheme in this procedure may improve the convergence rate.

REFERENCES

1. J. F. Thompson, Z.U.A. Warsi, and C.W. Mastin; "Numerical Grid Generation Foundations and Applications". North-Holland (1985).
2. A.S. Dvinsky and A.S. Popel; J. Comp. Phys. 67, 73 (1986).
3. L.W. Ehrlich; SIAM J. Sci. Stat. Comp. 7, 989 (1986).
4. Q.V. Dinh, R. Glowinski, B. Mantel, J. Periaux and P. Perrier; in "Computing Methods in Applied Sciences and Engineering", V. Edited by R. Glowinski and J.L. Lions. North-Holland, p. 123, (1982).
5. L. Fuchs; Computers & Fluids, 15, 69 (1986).
6. C-Y. Gu and L. Fuchs; in "Numerical Methods Laminar and Turbulent Flow-IV", Edited by C. Taylor et. al., Pineridge Press, p. 1501 (1985).
7. W.J. Usab and E.M. Murman; AIAA P-83-1946 (1983).
8. M.J. Berger; SIAM J. Sci. Stat. Comp. 7, 905 (1986).
9. L. Fuchs; "Proc. 7th GAMM conf. on Numerical Meth. in Fluid Dynamics", Vieweg Verlag, 20, p. 96 (1988).
10. A. Brandt; Math. Comp. 31, 331 (1977).
11. T. Thunell and L. Fuchs; in "Numerical Methods Laminar and Turbulent Flow-II". Edited by C. Taylor and B.A. Schrefler. Pineridge Press, p. 141 (1981).
12. L. Fuchs; in "Numerical Methods for Fluid Dynamics-II", edited by K.W. Morton and M.J. Baines, Clarendon Press, Oxford, p. 569, (1986).
13. W. Cherdron, F. Durst and J.H. Whitelaw; J. Fluid Mech. 84, 13, (1978).

Two-Dimensional Wind Flow over Buildings

Peter Kaps

Institute of Mathematics and Geometry

University of Innsbruck

Technikerstr. 13, A-6020 Innsbruck, Austria

Paper presented at the Fifth GAMM Seminar Kiel

Numerical Treatment of the Navier-Stokes Equations

January 20 - 22, 1989

Abstract

The incompressible Navier-Stokes equations are solved in a two-dimensional domain which corresponds to a special building shape. The numerical method we use is the random vortex method developed by Chorin because the problem is characterized by a high Reynolds number. Streakline diagrams are presented to illustrate a typical wind flow.

1. Introduction

Mathematically we want to solve the incompressible Navier-Stokes equations

$$\frac{\partial \mathbf{u}}{\partial t} + \mathbf{u} \cdot \nabla \mathbf{u} = \frac{1}{Re}\Delta \mathbf{u} - \operatorname{grad} p \quad \text{in } \Omega \times]0,\infty[. \quad (1.1)$$
$$\operatorname{div} \mathbf{u} = 0$$

We would like to have a region $\Omega \subset \mathbf{R}^2$ like the upper part of:

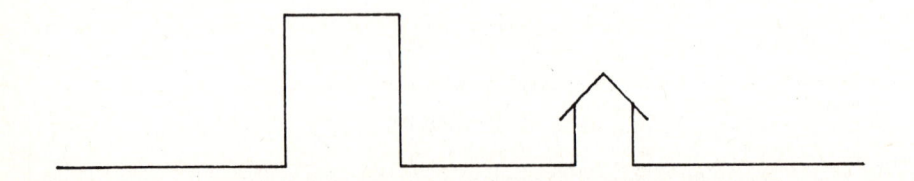

Until now we have the following form of Ω:

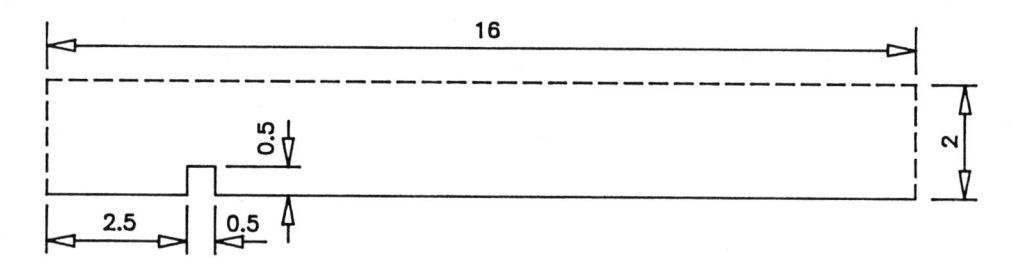

```
|←—————————————— 16 ——————————————→|
┌─────────────────────────────────────┐  ↑
│         0.5 ↓                       │  2
│    ┌──┐                             │  ↓
└────┘  └─────────────────────────────┘
|←2.5→|←0.5→|
```

The kinematic viscosity of air is given by, $\nu = 0.145 \times 10^{-4}$ m^2/sec. Thus, a typical length $L = 14.5$ m and a velocity $U = 1$ m/sec yield the Reynolds number $Re = 10^6$. For $x = 0$ we would like to prescribe the velocity profile for rural environment:

$$\mathbf{u} = (u, v), \qquad u = y^\alpha, \qquad v = 0, \qquad \alpha = 0.16. \tag{1.2}$$

A similar problem is studied in [9] and [12]. Since the Reynolds number of the problem is high we use the random vortex method developed by Chorin [4, 5, 6]. In this method, the concepts of vortex elements and vortex sheets play an essential role. In the following sections we give some motivation for the formulas we used in the computations. For more detailed information, see for example [7], [3] or [8]. Majda [10] gives a review on recent developments in vortex dynamics. In the presence of boundaries and viscosity, there is no convergence proof available. [13] present a numerical validation study of vortex methods. A recent study of the vortex sheet method is given in [11]. Finally, we mention the work of [1] and [2] on wind tunnels which was the origin of this paper.

Introducing the vorticity

$$\omega := \operatorname{curl} \mathbf{u} := \frac{\partial v}{\partial x} - \frac{\partial u}{\partial y} \tag{1.3}$$

the Navier-Stokes equations are given by

$$\frac{\partial \omega}{\partial t} + \mathbf{u} \cdot \nabla \omega = \frac{1}{Re} \Delta \omega. \tag{1.4}$$

We call the expression $\mathbf{u} \cdot \nabla \omega$ convection term and $\frac{1}{Re} \Delta \omega$ diffusion term.

Our aim is to compute from a given vorticity field $\omega(t)$ the velocity $\mathbf{u}(t)$ with help of $\operatorname{div} \mathbf{u} = 0$ and $\operatorname{curl} \mathbf{u} = \omega$. Then, (1.4) gives $\omega(t + \Delta t)$. We use two fractional steps: a convection step and a diffusion step.

2. Vortex elements

For the stream function ψ defined by

$$\mathbf{u} := \operatorname{curl} \psi := \left(\frac{\partial \psi}{\partial y}, -\frac{\partial \psi}{\partial x}\right) \tag{2.1}$$

it holds because of $\operatorname{curl} \mathbf{u} = \omega$:

$$-\Delta \psi = \omega. \tag{2.2}$$

$$G(\mathbf{x}) = -\frac{1}{2\pi} \log |\mathbf{x}| \tag{2.3}$$

is a fundamental solution, i.e. $-\Delta G = \delta$, where δ denotes the Dirac δ-function. G can be regarded as stream function of a point vortex.

$$\psi = G * \omega \tag{2.4}$$

is a solution of (2.2) since $-\Delta \psi = (-\Delta G) * \omega = \omega$. From (2.1) one obtains

$$\mathbf{u} = \mathbf{K} * \omega, \quad \mathbf{K} = \operatorname{curl} G = -\frac{(y, -x)}{2\pi |\mathbf{x}|^2}. \tag{2.5}$$

$\mathbf{K}$ has a singularity at $\mathbf{x} = 0$. To make $\mathbf{K}$ more regularly we follow Chorin and cut off $\mathbf{K}$ within a small disk of radius δ:

$$\mathbf{K}_\delta = \begin{cases} \mathbf{K} & |\mathbf{x}| \geq \delta \\ \mathbf{K} \cdot \dfrac{|\mathbf{x}|}{\delta} & 0 < |\mathbf{x}| < \delta \\ 0 & \mathbf{x} = 0. \end{cases} \tag{2.6}$$

The corresponding stream function is called **vortex blob** or **vortex element**:

$$f_\delta(\mathbf{x}) = -\frac{1}{2\pi} \begin{cases} \log |\mathbf{x}| & |\mathbf{x}| \geq \delta \\ \dfrac{|\mathbf{x}| - \delta}{\delta} + \log \delta & |\mathbf{x}| < \delta. \end{cases} \tag{2.7}$$

Approximating the stream function by N vortex elements of strength Γ_i at $\mathbf{x}_i$ we arrive at

$$\psi(\mathbf{x}) = \sum_{i=1}^{N} \Gamma_i f_\delta(\mathbf{x} - \mathbf{x}_i). \tag{2.8}$$

The velocity produced by these vortex elements is given by

$$\mathbf{u}_\delta(\mathbf{x}) = \sum_{i=1}^{N} \Gamma_i \mathbf{K}_\delta(\mathbf{x} - \mathbf{x}_i). \tag{2.9}$$

3. Motion of vortex elements

As already mentioned we use two fractional steps to solve (1.4).

i) **Convection step.** Here, we solve

$$\frac{\partial \omega}{\partial t} + \mathbf{u} \cdot \nabla \omega = 0. \tag{3.1}$$

$\mathbf{u}_\delta$ does not satisfy the potential boundary condition

$$\mathbf{u} \cdot \mathbf{n} = 0 \tag{3.2}$$

at solid boundaries. Therefore, we decompose $\mathbf{u}_\delta$ in the form $\mathbf{u}_\delta = \mathbf{u} + \operatorname{grad} q$, where $\operatorname{div} \mathbf{u} = 0$ and $\mathbf{u} \cdot \mathbf{n}|_{\partial \Omega} = 0$. With the potential velocity $\mathbf{u}_p = -\operatorname{grad} q$ it holds:

$$\mathbf{u} = \mathbf{u}_\delta + \mathbf{u}_p. \tag{3.3}$$

For the computation of $\mathbf{u}_p$ we use a stream function ψ_p. Since $\operatorname{curl} \mathbf{u}_p = 0$ and $\operatorname{div} \mathbf{u}_p = 0$, it holds

$$-\Delta \psi_p = 0. \tag{3.4}$$

We can introduce Dirichlet boundary conditions (except at the outflow). The normal boundary condition for $\mathbf{u}_p$ is known: $\mathbf{u}_p \cdot \mathbf{n} = -\mathbf{u}_\delta \cdot \mathbf{n}$. At the inflow we have to add (1.2). However, we cannot prescribe v. Since the tangent vector $\mathbf{t}$ is given by $\mathbf{t} = (n_y, -n_x)$, where $\mathbf{n} = (n_x, n_y)$ is the normal vector, it holds $\mathbf{u}_p \cdot \mathbf{n} = -\nabla \psi_p \cdot \mathbf{t}$ and one obtains ψ_p on the boundary by integration:

$$\psi_p(\mathbf{x}) = \psi_p(\mathbf{x}_0) + \int_{s_0}^{s} (\mathbf{u}_\delta \cdot \mathbf{n}) \, ds. \tag{3.5}$$

For convenience we use the SOR method to solve (3.4). However, this has two disadvantages: we have to introduce a grid although it is an advantage of the random vortex method to be grid free. Further, it is difficult to obtain higher derivatives of ψ_p whereas

$\mathbf{u}_\delta$ is analytical except at the center and the core radius of the vortex elements. At the grid points of the boundary we approximate (3.5) by

$$\psi_{k+1} = \psi_k + \Delta s (\mathbf{u}_\delta \cdot \mathbf{n})_k. \tag{3.6}$$

As outflow condition we prescribe $\dfrac{\partial \psi}{\partial x} = 0$.

Equation (3.1) states that ω is "transported" by the flow, i.e. ω is constant along the path lines

$$\frac{d\mathbf{x}_j}{dt} = \mathbf{u}(\mathbf{x}_j), \qquad j = 1, ..., N. \tag{3.7}$$

This is a large system of ODEs ($N \approx 1500$) which is not sparse. Since $\mathbf{u}$ is given by (3.3) an evaluation of the right hand side of (3.7) costs $(N-1)^2/2$ evaluations of $\mathbf{K}_\delta$. We solve (3.7) with the explicit Euler method:

$$\mathbf{x}_j(t + \Delta t) = \mathbf{x}_j(t) + \mathbf{u}(\mathbf{x}_j) \Delta t, \qquad j = 1, ..., N. \tag{3.8}$$

ii) **Diffusion step.** Here, we solve

$$\frac{\partial \omega}{\partial t} = \frac{1}{Re} \Delta \omega. \tag{3.9}$$

To simulate the transport of vorticity due to diffusion we add to the convective motion (3.8) a diffusive motion $\boldsymbol{\eta}_j = (\eta_x, \eta_y)_j$. To motivate the random distribution of $\boldsymbol{\eta}_j$ note that the Green's function for the one-dimensional form of the heat equation (i.e. $\Delta = \dfrac{\partial^2}{\partial x^2}$ in (3.9)) is given by $Gr(x,t) = \sqrt{\dfrac{Re}{4\pi t}} \exp\left(-\dfrac{Re}{4t} x^2\right)$. This corresponds to a Gauss distribution with zero mean and standard deviation $\sigma^2 = \dfrac{2t}{Re}$. In two dimensions the Green's function of (3.9) is given by $Gr(\mathbf{x},t) = Gr(x,t) Gr(y,t)$. Thus, we arrive at

$$\mathbf{x}_j(t + \Delta t) = \mathbf{x}_j(t) + \mathbf{u}(\mathbf{x}_j) \Delta t + \boldsymbol{\eta}_j. \tag{3.10}$$

The random motion $\boldsymbol{\eta}_j$ is given by $\boldsymbol{\eta}_j = (\eta_x, \eta_y)_j$, where η_x and η_y are $N(0, \sigma)$ distributed, independent Gaussian random numbers with

$$\sigma^2 = \frac{2\Delta t}{Re}. \tag{3.11}$$

4. Vortex sheets

Vortex sheets are introduced to satisfy the viscous boundary conditions $\mathbf{u} \cdot \mathbf{t} = 0$. Stopping the flow along the walls produces vorticity. We divide the walls into pieces of length S. The circulation is then given by

$$\Gamma = \mp \int_S (\mathbf{u} \cdot \mathbf{t})\, ds. \qquad (4.1)$$

At moderate and high values of the Reynolds number, it holds

$$\left|\frac{\partial v}{\partial x}\right| \ll \left|\frac{\partial u}{\partial y}\right| \qquad (4.2)$$

within a thin layer, the shear layer. Here, a coordinate system is used where x is parallel and y is vertical to the wall. The order of magnitude of this shear layer is

$$\Delta_s = O\left(\sqrt{\Delta t/Re}\right).$$

In the shear layer, it holds for the vorticity

$$\omega = -\frac{\partial u}{\partial y}. \qquad (4.3)$$

Replacing in (4.3) ω by the Dirac δ-function, we obtain for a sheet of finite length S similarly as for blobs:

$$K_\Delta(\mathbf{x}) = -H(y)\left\{H\left(x+\frac{S}{2}\right) - H\left(x-\frac{S}{2}\right)\right\} \text{ and}$$

$$u_\Delta(\mathbf{x}) = \sum_{i=1}^{M} \gamma_i K_\Delta(\mathbf{x} - \mathbf{x}_i) \qquad (4.4)$$

$$\gamma_i = \mp(\Delta u)_i.$$

γ_i is the circulation per unit length carried by the i-th vortex sheet. $(\Delta u)_i$ is the velocity jump across the i-th sheet. The expression $\{H(x+\frac{S}{2}) - H(x-\frac{S}{2})\}$ is 1 inside $[-\frac{S}{2}, \frac{S}{2}]$ and 0 outside. If at a solid boundary the velocity jump is given by Δu we introduce $\left[\frac{\Delta u\, S}{\Gamma_{min}} + 0.5\right]$ new sheets. Γ_{min} is the minimum vorticity.

5. Motion of vortex sheets

To compute the velocity of a vortex sheet we take an average over the length of the sheet:

$$u(\mathbf{x}_j) = \frac{1}{S} \int_{x_j - \frac{S}{2}}^{x_j + \frac{S}{2}} u(x', y_j) \, dx'.$$

One obtains

$$u(\mathbf{x}_j) = \sum_{i=1}^{M} \gamma_i \, D_{ij} \, H(y_j - y_i), \quad \text{where}$$

$$D_{ij} = \begin{cases} 1 - |x_i - x_j|/S & \text{if } |x_j - x_i| < S \\ 0 & \text{else.} \end{cases} \tag{5.1}$$

Note, that a sheet at $\mathbf{x}_i$ does not influence the sheet at $\mathbf{x}_j$ if $|x_j - x_i| > S$ or $y_j < y_i$. The y-component of the velocity is computed by integrating the continuity equation:

$$v(\mathbf{x}) = -\frac{\partial}{\partial x} \int_0^y u(x, y') \, dy'. \tag{5.2}$$

For a sheet at $\mathbf{x}_j$ we obtain

$$v(\mathbf{x}_j) = -(I^+ - I^-)/S \tag{5.3}$$

where $I^\pm$ denotes the integral in (5.2) evaluated at $(x_j \pm \frac{S}{2}, y_j)$. Vortex sheets diffuse perpendicular to the wall only. Thus, the motion of vortex sheets is given by

$$\mathbf{x}_j(t + \Delta t) = \mathbf{x}_j(t) + \mathbf{u}(x_j)\Delta t + (0, \eta)_j. \tag{5.4}$$

The components of $\mathbf{u}$ are given by (5.1) and (5.3).

For the transformation of vortex sheets into vortex blobs and vice versa see [7]. We just note that a vortex sheet is transformed into a vortex element with

$$\Gamma_j = S\gamma_j \tag{5.5}$$

if it is moved outside of the numerical boundary layer. Since sheets and blobs are two different representations of the same object they should produce the same field close to the wall. This gives a relation between core radius and sheet length:

$$\delta = S/\pi. \tag{5.6}$$

With these preparations we can describe the algorithm we used for our computations.

6. Algorithm

Start: N vortex elements, M vortex sheets given at time T

1. Compute $\mathbf{u}_p$.

   Compute boundary conditions with (3.6).

   Solve $-\Delta\psi_p = 0$ in $\Omega' = \Omega \cap (]0,16[\times]0,2[)$ with SOR.

2. Introduce vortex sheets to satisfy $\mathbf{u}\cdot\mathbf{t} = 0$ at solid walls.

3. Move vortex sheets according to (5.4).

4. Transforme sheets to blobs outside the numerical shear layer.

5. Move vortex blobs according to (3.10).

6. Transform blobs to sheets in numerical shear layer, appropriate treatment outside of Ω'.

7. $T = T + \Delta T$, go to 1.

We chose $S = 0.2$, $\Gamma_{min} = 0.05$, $\Delta t = 0.05$, and $h = 0.05$ as grid size for the SOR method.

7. Numerical results

We have executed the above algorithm at the Cyber 840 of the Computing Centre of the University of Innsbruck. The algorithm is time consuming. For example, to advance from $T = 12$ to $T = 13$ costs more than one hour computing time. We have performed computations for various Reynolds numbers leaving the other parameters unchanged. For $Re = 10^4$ and $Re = 10^5$ the algorithm stopped with the message "arithmetic indefinite". For $Re = 10^6$ and $Re = 10^7$ the algorithm worked. We give the results for $Re = 10^6$. We show that section of Ω with $0 \le x \le 8$ and $0 \le y \le 1$. The scale in x- and y-direction is different.

In Figure 1, we present plots of the motion of the vortex elements. We show the development of the flow for three different times, $T = 4, 8, 12$. The arrows indicate the direction of the flow and their length the magnitude of the velocity. The plots show a qualitatively correct representation of the flow.

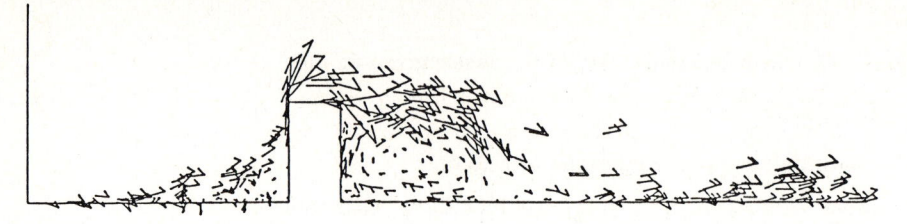

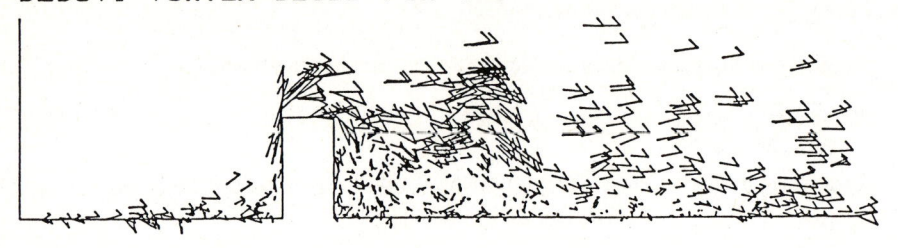

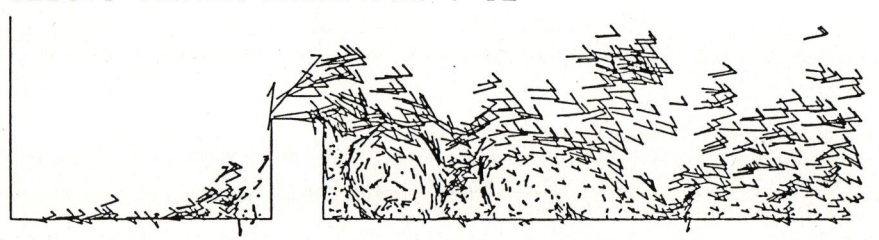

Figure 1. Motion of vortex elements

The next plot shows an average velocity profile for $T = 13$. The average is taken over 20 time steps. Obviously, the inflow condition (1.2) is not satisfied.

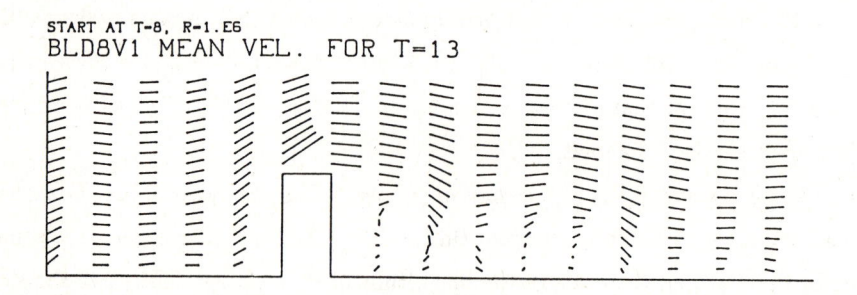

Figure 2. Average velocity profile for $T = 13$

8. References

[1] BUI, T.D.: *A Random Vortex Method for Flows at High Reynolds Number in a Tunnel: Applications to the Simulation of Wind Effects on Model Buildings.* Dept. of Computer Science, Concordia University, Montreal, Canada.

[2] BUI, T.D., AND HSIAO, C.C.: *Numerical Solution of the Navier-Stokes Equation for the Simulation of Wind Flow around Buildings.* Dept. of Computer Science, Concordia University, Montreal, Canada.

[3] CHEER, A.Y.: *Numerical Study of Incompressible Slightly Viscous Flow Past Blunt Bodies and Airfoils.* SIAM J. Sci. Stat. Comput. 4, 685-705 (1983)

[4] CHORIN, A.J.: *Numerical Study of Slightly Viscous Flow.* J. Fluid Mech. 57, 785-796 (1973)

[5] CHORIN, A.J.: *Vortex Sheet Approximation of Boundary Layers.* J. Comp. Phys. 27, 428-442 (1978)

[6] CHORIN, A.J., MARSDEN, J.E.: *A Mathematical Introduction to Fluid Mechanics.* Springer, New York (1984)

[7] GHONIEM, A.F., CHORIN, A.J., AND OPPENHEIM, A.K.: *Numerical Modeling of Turbulent Flow in a Combustion Tunnel.* Phil.Trans. R. Soc. Lond. A 304, 303-325 (1982)

[8] GHONIEM, A.F., GAGNON, Y.: *Numerical Investigations of Recirculating Flow at Moderate Reynolds Numbers.* AIAA-86-0370 (1986)

[9] HANSON, T., SUMMERS, D.M., AND WILSON, C.B.: *Numerical Modelling of Wind Flow over Buildings in Two Dimensions.* J. Num. Meth. in Fluids 4, 25-41 (1984)

[10] MAJDA, A.: *Vortex Dynamics: Numerical Analysis, Scientific Computing, and Mathematical Theory.* In ICIAM'87: Proceedings of the First International Conference on Industrial and Applied Mathematics, SIAM, Philadelphia (1988)

[11] PUCKETT, E.G.: *A Study of the Vortex Sheet Method and its Rate of Convergence.* SIAM J. Sci. Stat. Comput. 10, 298-327 (1989)

[12] SUMMERS, D.M., HANSON, T., AND WILSON, C.B.: *A Random Vortex Simulation of Wind-Flow over a Building.* J. Num. Meth. in Fluids 5, 849-871 (1985)

[13] SETHIAN, J., AND GHONIEM, A.: *Validation Study of Vortex Methods.* Submitted to J. Comp. Phys.

Laminar Shock/Boundary-Layer Interaction – A Numerical Test Problem

E. Katzer

Institute of Informatics and Applied Mathematics,
Christian-Albrechts University, D-2300 KIEL, FRG.

Abstract

The interaction of an oblique shock with a laminar boundary layer has been used frequently as a test problem for comparing Navier-Stokes codes. This paper identifies various length scales of the interaction region which impose cosiderable restrictions on the numerical grid. It is shown that the wall pressure distribution is not useful for assessing the numerical accuracy. The pressure at the edge of the boundary layer is proposed as an expedient variable indicating the numerical accuracy.

Local and global properties of the flow are analysed. A new similarity law for the length of the separation bubble facilitates an estimate of the global length scale and could be used for the generation of initial values for numerical calculations.

Introduction

Shock/boundary-layer interaction is a typical phenomenon of viscous - inviscid interaction and is of particular importance in aerodynamics. The interaction with a laminar boundary layer is relevant for the flow through turbomachines (figure 1) and for hypersonic flows around space vehicles.

The interaction of an oblique shock with a laminar boundary layer on a flat plate has been used frequently as a test problem for verifying Navier-Stokes calculations. The present paper discusses the significance of the grid for the numerical resolution. The necessary step sizes of the grid required for accurate solutions will be estimated and parameters for assessing the accurracy are proposed. Before discussing numerical aspects, we recall the main physical properties of laminar interaction.

First recall the inviscid case. A shock impinging on a wall is reflected as a second shock. The pressure increase across both shocks, p_3/p_1, measures the shock strength. In the viscous case, figures 2 and 3, the boundary layer is retarded by the shock. It separates and brakes away from the wall. The impinging shock is reflected as an expansion fan, thereby turning the shear layer towards the wall where the flow reattaches. Displacement effects of the boundary layer generate two compression zones emanating from the separation and reattachment region enclosing a constant pressure plateau. A surprising property of laminar interaction is the large extent of the separation bubble (Liepmann et al. [2]). (Notice that the scales normal to the wall are considerably stretched in figures 2 to 7.)

The region in the vicinity of the separation point is called the "free interaction region" (Chapman et al. [3]). The flow properties there are determined by a local mechanism which was first described by Oswatitsch [4]. A comprehensive description of the free interaction region is given by the "triple deck", an asymptotic theory, valid for high Reynolds numbers, Re. Main result of this theory is the formation of a "lower deck" at the wall with length scale of order $Re^{-3/8}$ and height of order $Re^{-5/8}$ which displaces the main part of the boundary layer (Stewartson [5]).

Solution of the Navier-Stokes equations

The flow fied is governed by the Navier-Stokes equations which yeald for a control volume Vol :

$$\frac{\partial}{\partial t}\int_{Vol}\rho\, dV = -\int_{\partial Vol}\rho\,(vds) \tag{1}$$

$$\frac{\partial}{\partial t}\int_{Vol}\rho v\, dV = -\int_{\partial Vol}\rho v\,(vds) + \int_{\partial Vol} S\, ds \tag{2}$$

$$\frac{\partial}{\partial t}\int_{Vol}\rho e_t\, dV = -\int_{\partial Vol}\rho e_t\,(vds) + \int_{\partial Vol}(Sv - q)\, ds \tag{3}$$

where $\rho, \rho v, \rho e_t$ are the density, momentum, and total energy per unit volume; S is the stress tensor of a Newtonian fluid and q is Fourier's law of heat conduction. The fluid is assumed to be a perfect gas with viscosity obeying Sutherland's law and Stokes' hypothesis.

The boundary condition at the inflow boundary, located some distance downstream of the leading edge, is specified by an undisturbed flat plate boundary layer. Vanishing gradients:

$$\frac{\partial}{\partial x}(\rho, \rho v, \rho e_t) = \frac{\partial^2}{\partial x^2}\rho u = 0 \tag{4}$$

are assumed at the outflow boundary. The no-slip boundary condition is applied at the adiabatic wall together with a vanishing pressure gradient $\partial p/\partial y = 0$. At the farfield boundary, undisturbed flow is specified upstream of the impinging shock and Rankine-Hugoniot values are prescribed downstream.

The governing equations (1-3) are solved using a finite volume version of the explicit time-split MacCormack scheme [6], [7]. Stress tensor and heat flux are approximated by Deiwert's [8] approach. Details are presented elsewhere [9]. Starting from the inviscid solution, the numerical calculation proceeds in time until a steady state is reached.

Figure 2 shows a typical result for shocks strong enough to provoke separation and compares the numerical solution with the experiment of Greber, Hakkinen et al. [10], [11]. The numerical flow parameters are: $M_1 = 2.0$, $Re_{x0} = 3 \cdot 10^5$, $p_3/p_1 = 1.4$. The boundary layer profiles and the wall pressure agree well with the experiment. The shape of the wall shear stress distribution corresponds with the experiment but the length of the separation bubble seems to be overestimated by the calculation. This is a surprising result if one considers the good match in the wall pressure and it is suspected that the experiment is distorted by the Stanton probe which has been used to determine the wall shear stress.

The wall shear stress exhibits a second minimum just upstream of the reattachment point. The author has found this second minimum in all his numerical calculations, except for very small separation bubbles. The reason for this second minimum can be seen in figure 4 which shows the streamlines in the separation bubble. (Note that these streamlines do not represent equidistant values of the stream function.) The streamlines reveal that the centre of the bubble with the fluid at rest is shifted to the downstream flank of the separation bubble. Displacement effects of the centre accelerate the reverse flow, thereby causing the second minimum.

The streamlines in the oncoming part of the boundary layer show no distortions when the fluid passes the separation bubble. The main part of the boundary layer is just displaced by the separation bubble. This behaviour corresponds with results of the triple-deck theory. The shape of the boundary layer profiles, figure 2, and the distribution of the boundary-layer dispacement thickness, δ^* and the momentum thickness, δ^{**}, (figure 4) confirm this result. The slight increase in the momentum thickness is the same as that for an undisturbed boundary layer, whereas the displacement thickness follows the shape of the separation bubble. This is typical for a boundary layer which is displaced from the wall by fluid with small velocity.

Influence of the grid and numerical accuracy

Various physical length scales of the shock/boundary-layer interaction impose substantial restrictions on the numerical grid. A reliable numerical solution requires that the grid spacings are an order of magnitude smaler than these length scales. Therefore, nonequidistant grids are applied. Figure 5 shows the grid consisting of 151 * 101 mesh points used for the calculation of the results in figure 2. Notice that only every second grid line is plotted.

An adequate resolution of the boundary layer requires that the spacings normal to the wall are smaller than the thickness of the boundary layer:

$$\Delta y \ll \delta^* \sim Re^{-1/2}. \tag{5}$$

The resolution of the triple-deck structure impose more restrictive conditions on the grid spacings near separation and reatachment point. The lower-deck length scale requires:

$$\Delta y \ll \delta^* \cdot Re^{-1/8} \sim Re^{-5/8} \tag{6}$$
$$\Delta x \approx \Delta y \cdot Re^{1/4}. \tag{7}$$

Figures 6 and 7 show the influence of the grid spacings on the numerical solution. Obviously the coarse grid does not resolve the interaction region accurately. The impinging shock and the compression and expansion zones are spread over a considerable width and the length and shape of the separation bubble is distorted. In contrast, the wall pressure curves are in close correspondance with each other. This shows, that a comparison of the wall pressure with experimental data is not sufficient for assessing the numerical accuracy. Therefore, the author proposes an inspection of the pressure at the edge of the boundary layer.

Figure 8 compares the pressure at the edge of the boundary layer for different grids. The numerical dissipation parameter, ϵ, is also varied, but further investigations showed, that the numerical dissipation damps numerical oscillations without distorting the solution in the present parameter range. With the coarse grid (upper diagram), the boundary layer is subject to a gradual pressure gradient instead of a discontinuity. This reduces the bubble length. The finest grid (lower diagram) with small dissipation gives sharp pressure gradients but shows small numerical oscillations. A redistribution of the grid with reduced grid spacings near shock impingement (center diagram), together with increased numerical dissipation, prevents these oscillations. The shock width is only slightly increased and the bubble length is not changed. These results show, that the numerical width of the impinging shock should be an order of magnitude smaller, than the extent of the interaction region. Near shock impingement, small grid spacings are essential not only normal to the wall but also in longitudinal direction. A similar result has been obtained by Messina [12].

It should be noted, that numerical oscillations may not be detected in the wall pressure distribution. This is not surprising because even the shock is not visible in the wall pressure.

Local and global scaling laws

The pressure at the separation point in the free interaction region is governed by a local scaling law:

$$c_{p_s} = P_s \sqrt{c_{f_1}/(M_1^2 - 1)^{1/2}} \;, \tag{8}$$

where the skin friction at the beginning of the interaction region is:

$$c_{f_1} = \frac{0.664}{\sqrt{Re_{x1}/C}} \;. \tag{9}$$

The Chapman-Rubesin constant $C = T_\infty \mu_w / T_w \mu_\infty$ is the ratio of free stream values of viscosity and temperature to wall values and M_1 is the free stream Mach number. Figure 9 shows, that the author's results in the range $1.4 < M_1 \leq 3.4$, $10^5 \leq Re_{x_0} \leq 6 \cdot 10^5$ and for shock strength of $1.2 \leq p_3/p_1 \leq 1.8$ confirm equation (8) with a constant of $P_s = 1.4$. The same scaling law is valid for the plateau pressure for well established plateaus with a constant of $P_p = 2.3$.

Further numerical investigations (Katzer [13]) confirme, that the length scale of the free interaction region is independent of the shock strength. The length scales of the triple deck could not be verified, because the Reynolds numbers are much too low for application of an asymptotic theory.

Interpolation of the numerical data obtained for the length of the separation bubble leads to a new similarity law for this global length scale:

$$\frac{l_B}{\delta_0^*} \frac{M_1^3}{\sqrt{Re_{x_0}/C}} = 4.4 \, \frac{p_3 - p_{inc}}{p_1} \;. \tag{10}$$

Here δ_0^* is the displacement thickness of an undisturbed flat plate boundary layer at position x_0 where the shock impinges at the wall. The Reynolds number Re_{x_0} is based on this length and on free stream values. The pressure for incipient separation is given by equation (8) using the undisturbed skin friction at position x_0 and a constant of $P_{inc} = 2.62$. Figure 10 shows that the numerical data are well interpolated by equation (10) in the range of the present investigation. The reader is referred to [9] for a comparison of other characteristic flow variables, e. g. the separation angle.

Conclusion

Various physical length scales of shock/boundary-layer interaction have been identified, namely:

- the extent of the interaction region,
- the boundary-layer thickness,
- the size of the lower deck and
- the numerical width of the shock.

These length scales impose several restrictions on the numerical grid. Small grid spacings are required along the wall and in orthogonal direction for an accurate numerical resolution. An inspection of the pressure distribution at the edge of the boundary layer is proposed for assesing the numerical accuracy. The numerical width of the impinging shock should be an order of magnitude smaller than the extent of the interaction region.

Several scaling laws concerning local and global properties of the interaction region are investigated. The pressure at the separation point represents a local scaling law of the free interaction region. A new similarity law for the length of the separation bubble constitutes a global length scale. The displacement property of the separated region together with the bubble length could be used for the construction of initial conditions for numerical calculations.

Acknowledgement

The present work was carried out at the Institute of Theoretical Fluid Mechanics of the German Aerospace Research Establishment (DLR-AVA) in Göttingen, FRG. My particular thanks are due to Prof. H. Oertel and Prof. J. Zierep for supporting the work and for many fruitful discussions.

References

[1] Graham, C.G.; Kost, F.H.:
Shock boundary layer interaction on high turning transonic tubine cascades.
ASME Publ. 79-GT-37, 1979.

[2] Liepmann, H.W.; Rhosko, A.; Dhawan, S.:
On reflection of shock waves from boundary layers.
NACA Report 1100, 1952.

[3] Chapman, D.R.; Kuehn, D. M.; Larson, H.K.:
Investigation of separated flows in supersonic and subsonic streams with emphasis on the effect of transition.
NACA Report 1356, 1958.

[4] Oswatitsch, K; Wieghardt, K:
Theoretische Untersuchungen über stationäre Potentialströmungen und Grenzschichten bei hohen Geschwindigkeiten.
Lilienthal Bericht S13 / 1. Teil, 1942, pp. 7-24. English translation:
Theoretical analysis of stationary potential flows and boundary layers at high speed. NACA TM 1189, 1948.

[5] Stewartson, K.:
Multistructured boundary layers on flat plates and related bodies.
Advances in Applied Mechanics 14 (1974), pp. 145-239.

[6] MacCormack, R.W.:
The effect of viscosity in hypervelocity impact cratering.
AIAA Paper 69-354, 1969.

[7] MacCormack, R.W.; Baldwin, B.S.:
A numerical method for solving the Navier-Stokes equations with application to shock-boundary layer interactions.
AIAA Paper 75-1, 1975.

[8] Deiwert, G.S.:
Numerical simulation of high Reynolds number transonic flows.
AIAA J. 13 (1975), pp. 1354-1359.

[9] Katzer, E.:
Numerische Untersuchung der laminaren Stoß-Grenzschicht-Wechselwirkung.
Thesis, Karlsruhe 1985, DFVLR-FB 85-34. English translation:
Numerical study of laminar shock/boundary layer interaction. ESA-TT-958, 1986.

[10] Greber, I.; Hakkinen, R.J.; Trilling, L.:
Laminar boundary layer oblique shock wave interaction on flat and curved plates.
ZAMP 9b (1958), pp. 312-331.

[11] Hakkinen, R.J.; Greber, I.; Trilling, L.; Abarbanel, S.S.:
The interaction of an oblique shock wave with a laminar boundary layer.
NASA Memo 2-18-59W, 1959.

[12] Messina, N.A.:
A computational investigation of shock waves, laminar boundary layers and their mutual interaction.
Ph. D. Thesis, Dept. of Aerospace and Mechanical Sci., Princeton, 1977.

[13] Katzer, E.:
On the length scales of laminar shock/boundary-layer interaction.
J. Fluid Mech. 206 (1989), pp. 477-496.

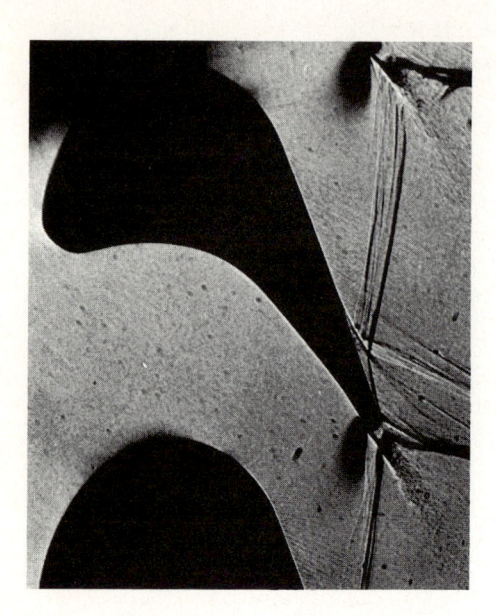

Fig. 1: Laminar shock/boundary-layer interaction in turbine cascades (Graham and Kost [1]).

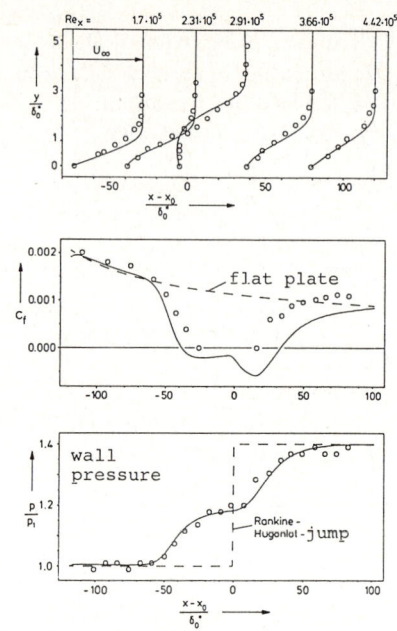

Fig. 2: Comparison of the numerical solution with the experiment.
———: Navier-Stokes solution
oooo: Experiment, Hakkinen

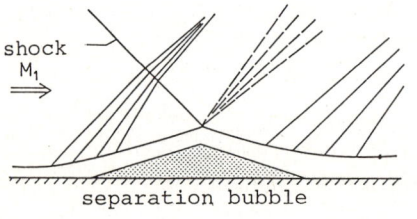

Fig. 3: Sketch of interaction region.

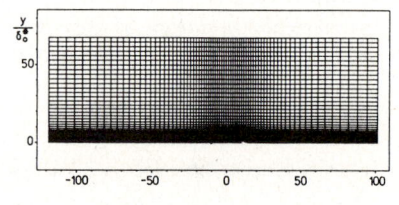

Fig. 5: Numerical grid.

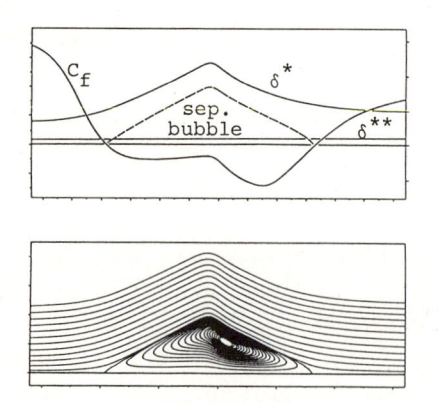

Fig. 4: Streamlines in boundary layer and separation bubble.

Fig. 6: Grid influence on inviscid flow field.

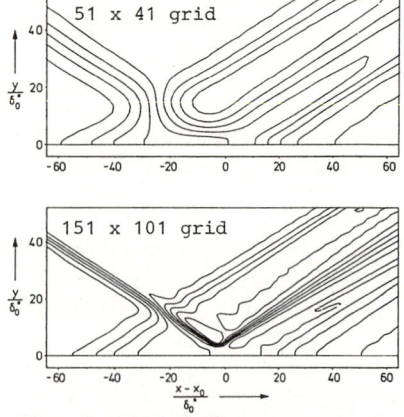

Fig. 7: Influence of grid size.

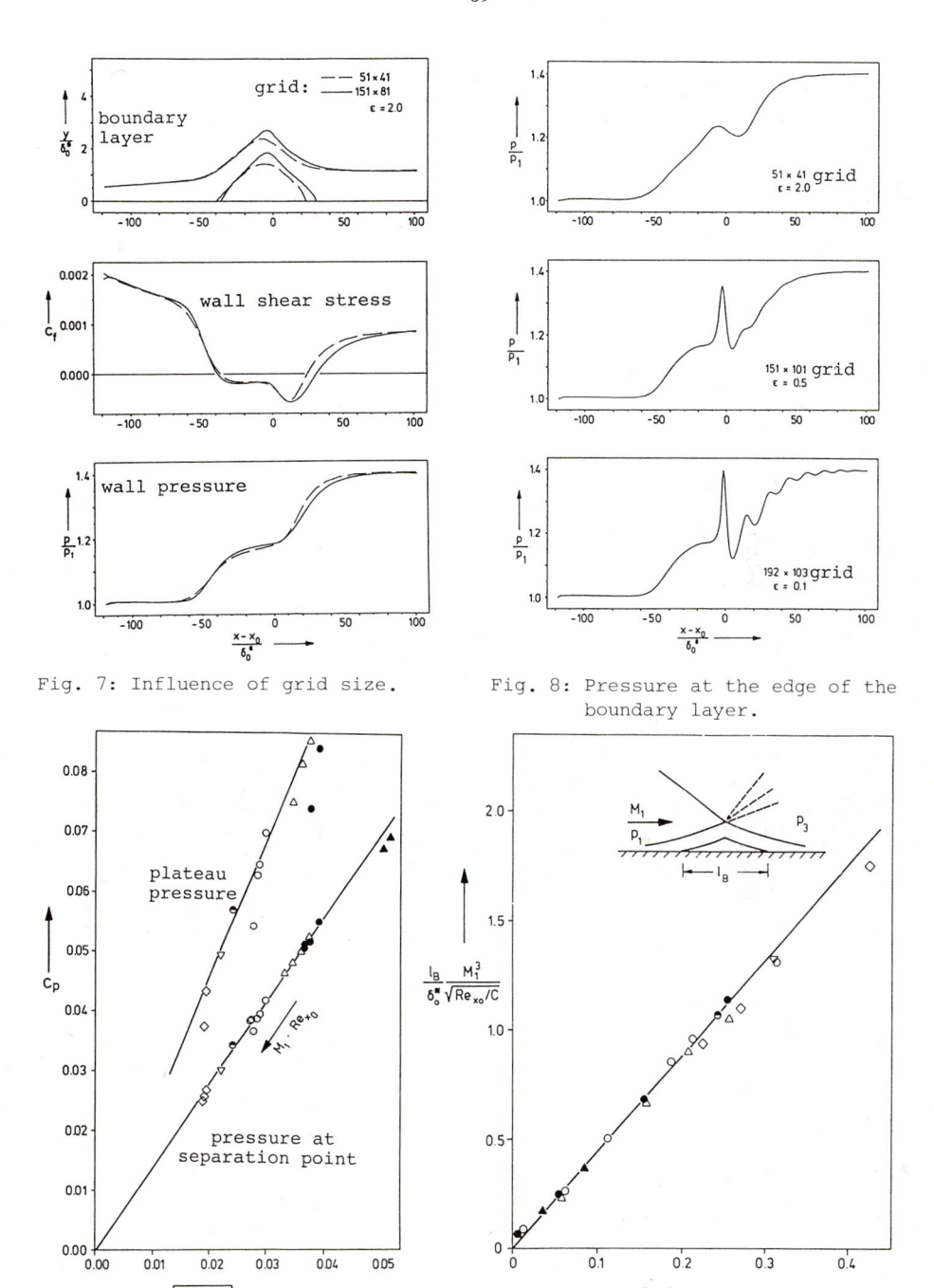

Fig. 8: Pressure at the edge of the boundary layer.

Fig. 9: Pressure at the separation point and plateau pressure.

Fig. 10: Length of the separation bubble.

COMPARISON OF UPWIND AND CENTRAL FINITE-DIFFERENCE METHODS FOR THE COMPRESSIBLE NAVIER-STOKES EQUATIONS

B. Müller
DLR Institut für Theoretische Strömungsmechanik
Bunsenstraße 10, D-3400 Göttingen, Federal Republic of Germany

SUMMARY

Two implicit second-order finite-difference methods are compared for the steady-state solution of the time-dependent compressible Navier-Stokes equations: a central spatial discretization scheme with added second- and fourth-order numerical damping and an upwind scheme, which reduces to first-order accuracy at extrema and is total variation diminishing for nonlinear one-dimensional scalar hyperbolic equations.

The upwind scheme proves in general to be more robust and more accurate than the central scheme for subsonic flat plate flow, transonic airfoil flow, and hypersonic ramp flow. Using approximate factorization, the central scheme is more efficient than the upwind scheme.

INTRODUCTION

Second-order central spatial differencing methods have been widely used to solve the compressible Navier-Stokes equations (cf. e.g. [1]). Similar to central schemes for the compressible Euler equations, numerical damping is usually introduced by adding blended second and fourth differences in the dependent variables (cf. e.g. [2]).

During the past decade, high-resolution upwind schemes for the Euler equations have demonstrated an improvement in robustness, accuracy, and in connection with relaxation methods even efficiency (cf. e.g. [3]). Although upwind methods have been used to solve the Navier-Stokes equations with the viscous fluxes in general second-order central differenced, their accuracy has been investigated only recently (cf. e.g. [4, 5, 6]). Whereas flux vector splittings should be modified for viscous flow simulations [5], no necessary changes have been reported for flux difference splittings to maintain accuracy in shear layers (cf. e.g. [7]).

The objective of the present paper is to compare robustness, accuracy, and efficiency of a second-order accurate central difference scheme and a high-resolution upwind method based on flux difference splitting for subsonic flow over a flat plate, transonic flow over a NACA 0012 airfoil, and hypersonic flow over a 15° ramp. The central and upwind space discretization methods employ implicit approximate factorization time-stepping, the same initial and boundary conditions, and the same meshes.

COMPRESSIBLE NAVIER-STOKES EQUATIONS

For two-dimensional Newtonian fluid flow of perfect gas, the time-dependent compressible Navier-Stokes equations read in dimensionless conservation-law form and in general coordinates as follows (cf. e.g. [8, 9]):

$$\frac{\partial q}{\partial \tau} + \frac{\partial E^{(\xi)}}{\partial \xi} + \frac{\partial E^{(\eta)}}{\partial \eta} = Re_\infty^{-1} \left[\frac{\partial E_v^{(\xi)}}{\partial \xi} + \frac{\partial E_v^{(\eta)}}{\partial \eta} \right] \qquad (1)$$

where $q = J^{-1}(\rho, \rho u, \rho v, e)^T$ is the vector of the conservative variables scaled by the Jacobian J of the transformation of the independent variables

$$\tau = t, \; \xi = \xi(t, x, y), \; \eta = \eta(t, x, y). \qquad (2)$$

$E^{(k)}$ and $E_v^{(k)}$ are the inviscid and viscous fluxes, resp., in the k-direction. The freestream Reynolds, Prandtl, and Mach numbers Re_∞, Pr_∞ and M_∞, resp., are defined in terms of the freestream values denoted by subscript ∞ and a characteristic length L. Stokes's hypothesis and usually the Sutherland law are employed. The Prandtl number and the ratio of specific heats are constant, namely Pr = 0.72 and γ = 1.4.

Ordering the viscous terms according to their derivatives

$$E_v^{(\xi)} = F_1\left(q, \frac{\partial q}{\partial \xi}\right) + F_2\left(q, \frac{\partial q}{\partial \eta}\right),$$

$$E_v^{(\eta)} = G_1\left(q, \frac{\partial q}{\partial \xi}\right) + G_2\left(q, \frac{\partial q}{\partial \eta}\right), \qquad (3)$$

the thin-layer approximation amounts to neglecting $E_v^{(\xi)}$ and G_1. Employing the thin-layer equations assumes that the body contour conforms to a line of constant η and that among the viscous terms only the gradients in the near-normal η-direction need to be resolved.

CENTRAL FINITE-DIFFERENCE METHOD

The compressible Navier-Stokes equations (1) are solved at the interior grid points of a boundary-fitted structured mesh. The approximate factorization scheme of Beam and Warming [1] is employed as basic implicit algorithm:

$$\left[I + \frac{\vartheta \Delta \tau}{1+\psi}\left(\mu_\xi \delta_\xi \frac{\partial E^{(\xi)n}}{\partial q} - Re_\infty^{-1} \delta_\xi \frac{\partial F_1^n}{\partial q}\right) + D_I^{(\xi)n} \right] \Delta q^{*n} =$$

$$= -\frac{\Delta \tau}{1+\psi} \left[\mu_\xi \delta_\xi E^{(\xi)n} + \mu_\eta \delta_\eta E^{(\eta)n} - \right. \qquad (4a)$$

$$\left. - Re_\infty^{-1}\left(\delta_\xi F_1^n + \mu_\xi \delta_\xi F_2(\tilde{q}^n) + \mu_\eta \delta_\eta G_1(\tilde{q}^n) + \delta_\eta G_2^n\right) \right] + D_E^n ,$$

$$\left[I + \frac{\vartheta \Delta \tau}{1+\psi}\left(\mu_\eta \delta_\eta \frac{\partial E^{(\eta)n}}{\partial q} - Re_\infty^{-1} \delta_\eta \frac{\partial G_2^n}{\partial q}\right) + D_I^{(\eta)n} \right] \Delta q^n = \Delta q^{*n} , \qquad (4b)$$

$$q^{n+1} = q^n + \Delta q^n \qquad (4c)$$

where $\tilde{q}^n = q^n + \vartheta \Delta q^{n-1}$.
I is the 4 x 4 identity matrix. $\Delta \tau$ denotes the time step, and n the time level.

The first-order Euler implicit time differencing formula, i.e. $\psi = 0$ and $\vartheta = 1$ in (4), is used for steady-state calculations. The second-order three-point-backward formula, i.e. $\psi = 1/2$ and $\vartheta = 1$ in (4), is preferred for unsteady flow simulations. The classical finite-difference operators are defined by

$$\delta_\xi a_{i,j} = a_{i+1/2,j} - a_{i-1/2,j} \quad , \quad \mu_\xi a_{i,j} = (a_{i+1/2,j} + a_{i-1/2,j})/2 \quad , \text{etc.} . \quad (5)$$

For convenience, $\Delta \xi = \Delta \eta = 1$ is assumed. The spatial discretizations are second-order accurate on a smooth mesh. The difference equations (4a) and (4b) lead to block-tridiagonal linear systems, which are solved by the Richtmyer algorithm.

If the explicit and implicit numerical damping terms D_E and $D_I^{(k)}$ in (4) were zero, the discretization would be perfectly central, but short wavelengths oscillations could hardly be damped for high Reynolds numbers. Therefore, numerical damping is introduced by nonlinear second and linear fourth differences in the conservative variables [9] similar to [2]:

$$D_E = \epsilon^{(\xi)} \delta_\xi (\mu_\xi \sigma) \delta_\xi (Jq) + \epsilon^{(\eta)} \delta_\eta (\mu_\eta \sigma) \delta_\eta (Jq) - \epsilon_E J^{-1} [\delta_\xi^4 + \delta_\eta^4](Jq) , \quad (6a)$$

$$D_I^{(k)} = - \epsilon^{(k)} \delta_k (\mu_k \sigma) \delta_k J - \epsilon_I J^{-1} \delta_k^2 J \quad (6b)$$

where

$$\epsilon_E = \Delta \tau \quad , \quad \epsilon_I = \begin{cases} 2 \epsilon_E & \text{for } \psi = 0 \text{ and } \vartheta \geq \dfrac{1}{2} , \\ 3.6 \epsilon_E & \text{for } \psi = \dfrac{1}{2} \text{ and } \vartheta = 1 , \end{cases}$$

$$\varepsilon_{i,j}^{(\xi)} = \kappa_2 \Delta \tau \frac{M_{i,j}}{M_\infty} \max \left(\gamma_{i+1,j}^{(\xi)} , \gamma_{i,j}^{(\xi)} , \gamma_{i-1,j}^{(\xi)} \right) , \text{etc.} ,$$

$$\sigma = J^{-1} \left[|W^{(\xi)}| + c |\nabla \xi| + |W^{(\eta)}| + c |\nabla \eta| \right]$$

with

$$\gamma_{i,j}^{(\xi)} = \left| \frac{p_{i+1,j} - 2p_{i,j} + p_{i-1,j}}{p_{i+1,j} + 2p_{i,j} + p_{i-1,j}} \right| \quad , \text{etc.} ,$$

$$M = \frac{|u|}{c} \quad , \quad \kappa_2 = \frac{1}{2} \quad , \quad W^{(k)} = k_t + k_x u + k_y v .$$

The explicit fourth-order differences are modified adjacent to the boundaries by assuming the second differences in Jq to be zero at the boundaries [8, 9]. Although fourth-order differences should be switched off across a shock [10], the present numerical damping model (6) has worked reasonably well even for hypersonic flow simulations [9, 11].

The implicit algorithm (4) with the numerical damping terms (6) will be denoted here as "central method", although the numerical damping model (6) introduces an upwind effect [10].

UPWIND FINITE-DIFFERENCE METHOD

The upwind scheme based on the modified flux approach of Harten and Yee [12] is used here. For one-dimensional hyperbolic systems of conservation laws with constant coefficients or with nonlinear fluxes in the scalar case, Harten showed that his high-resolution scheme is total variation diminishing (TVD) in the sense of non-increasing and second-order accurate except for extrema where the accuracy is reduced to first-order [13]. The upwind TVD scheme can be implemented into the basic implicit algorithm (4) by choosing [7, 14]:

$$D_E = -\frac{\Delta\tau}{1+\psi}\frac{1}{2}\left[\delta_\xi(\mu_\xi J)^{-1} R^{(\xi)}\phi^{(\xi)} + \delta_\eta(\mu_\eta J)^{-1} R^{(\eta)}\phi^{(\eta)}\right], \tag{7a}$$

$$D_I^{(k)} = -\frac{\vartheta\Delta\tau}{1+\psi}\frac{1}{2}\left[\delta_k(\mu_k J)^{-1}\lambda^{(k)} I \delta_k J\right] \tag{7b}$$

where the components of $\phi^{(\xi)}$ are defined by (supressing the j-index):

$$\phi_{i+1/2}^{(\xi)l} = Q\left(\lambda_{i+1/2}^{(\xi)l}\right)\mu_\xi g_{i+1/2}^{(\xi)l} - Q\left(\lambda_{i+1/2}^{(\xi)l} + \gamma_{i+1/2}^{(\xi)l}\right)\alpha_{i+1/2}^{(\xi)l} \tag{8}$$

with $\alpha_{i+1/2}^{(\xi)} = R_{i+1/2}^{(\xi)^{-1}}\delta_\xi(Jq)_{i+1/2}$,

$g_i^{(\xi)l} = \text{minmod}\left(\alpha_{i-1/2}^{(\xi)l}, \alpha_{i+1/2}^{(\xi)l}\right)$,

$\gamma_{i+1/2}^{(\xi)l} = \frac{1}{2} Q\left(\lambda_{i+1/2}^{(\xi)l}\right)\left(\delta_\xi g_{i+1/2}^{(\xi)l}\right)/\alpha_{i+1/2}^{(\xi)l}$ for $\alpha_{i+1/2}^{(\xi)l} \neq 0$.

Q denotes the entropy function [13], and $\lambda^{(\xi)l}$ the l-th eigenvalue of $\partial E^{(\xi)}/\partial q$. The columns of $R_{i+1/2}^{(\xi)}$ are the right eigenvectors of $\partial E^{(\xi)}/\partial q$ evaluated with the Roe average [3] and the arithmetically averaged metric terms. The second eigenvector corresponding to the eigenvalue of multiplicity two is scaled by the speed of sound c, because dimensional arguments suggest a velocity and c is strictly positive and easily available. The minmod limiter [12] is used with artificial compression applied to the linearly degenerate fields. Fictitious values of α outside the boundaries are obtained by zeroth-order extrapolation, e.g. $\alpha_{1/2}^{(\xi)} = \alpha_{1+1/2}^{(\xi)}$.

The entropy parameter in the definition of Q is chosen here as an anisotropic function of the spectral radii $\lambda^{(k)}$ of $\partial E^{(k)}/\partial q$ [15]:

$$\delta^{(\xi)} = \tilde{\delta}\lambda^{(\xi)}\left[1 + \left(\frac{\lambda^{(\eta)}}{\lambda^{(\xi)}}\right)^{\frac{2}{3}}\right] \tag{9}$$

with $\tilde{\delta} = \frac{1}{80}$ in general.

GRID GENERATION, BOUNDARY TREATMENT, INITIALIZATION

The grids are generated algebraically. For the high Reynolds number flow simulations, the grid points are clustered close to the walls (Figs. 1a, 3a). The C-grid around the NACA 0012 airfoil was generated by T. Berglind, FFA, The Aeronautical Research Institute of Sweden, using the transfinite interpolation method (Fig. 2a).

At a solid wall, the no-slip condition and either isothermal or adiabatic temperature conditions are imposed. The wall pressure is determined from the wall-normal momentum equation or from its boundary layer approximation for the ramp flow case. At the symmetry boundaries y = 0, v and the first y-derivatives of ρ, u, and p vanish. At the farfield boundaries, the locally one-dimensional Riemann invariants for isentropic flow, entropy, and tangential velocity component are given their freestream values, or extrapolated, if the corresponding characteristic is coming from outside or inside, resp., the considered domain. At the outflow boundary, all the conservative variables are linearly extrapolated, except for e if $M_\infty < 1$. In that case, the pressure is fixed to its freestream value.

The flowfields are initialized by freestream conditions, except for solid walls where the no-slip and isothermal temperature conditions are imposed from the very beginning.

RESULTS

SUBSONIC FLAT PLATE FLOW

The 49 x 17 H-mesh in Fig. 1a is employed for the simulation of laminar flow over an adiabatic semi-infinite flat plate at $M_\infty = 0.5$ and $Re_\infty = 2 \times 10^5$ based on the distance L between the leading edge and the outflow boundary. The local time-stepping option [2]

$$\Delta \tau = \frac{\Delta \tau_{ref}}{1 + \sqrt{J}} \tag{10}$$

is chosen with $\Delta\tau_{ref} = 1$. Whereas $\Delta\tau_{ref} \geq 2$ leads to negative values of the temperature for the central scheme, even $\Delta\tau_{ref} = 5$ can be safely taken for the upwind scheme.

The thin-layer approximation is used, because the solution of the full Navier-Stokes equations takes about 30 % more CPU time without significant deviations from the thin-layer results.

The central scheme converges faster than the upwind scheme: after 800 time levels, max $\{\|\Delta\rho\|_2, \|\Delta\rho u\|_2, \|\Delta\rho v\|_2, \|\Delta e\|_2\} = 10^{-6}$ for the central scheme and 60 x 10^{-6} for the upwind scheme. The CPU time on the IBM 3090 computer of the DLR without using the vector processor option is 196 and 256 seconds for the central and upwind schemes, respectively.

The accuracy of the results on the coarse mesh (Fig. 1a) is assessed by comparison with the compressible boundary layer solution of Chapman and Rubesin. The skin friction coefficient predicted by the central scheme is closer to the Chapman-Rubesin solution than the result of the upwind scheme (Fig. 1b), because the central method determines the velocity component u near the wall more accurately (Fig. 1c) and the temperature and thereby the viscosity at the wall larger than the upwind method (cf. Fig. 1e). Thus, the upwind scheme appears to be slightly too anti-diffusive near the wall. However, near the boundary layer edge, the upwind scheme is more accurate than the central scheme. The velocity component v is predicted more accurately by the upwind scheme, too (Fig. 1d). Whereas the central method yields a converged wall temperature that is too large (Fig. 1e), the upwind method has not yet converged and approaches the boundary layer solution when the number of time levels is increased.

Mesh refinement leads to close agreement of the results predicted by the central and upwind schemes. Deviations from the Chapman-Rubesin solution, namely $u/u_\infty > 1$ at the boundary layer edge and $v/u_\infty \to 0$ for $y \to \infty$ for the Navier-Stokes solution, are consequences of the boundary layer approximation of the y-momentum equation.

TRANSONIC AIRFOIL FLOW

The test case A2 of the GAMM Workshop on "Numerical Simulation of Compressible Navier-Stokes Flows" held in Nice in 1985 is considered: laminar flow over a NACA 0012 airfoil at $M_\infty = 0.8$, $\alpha = 10°$, $Re_\infty = 500$ (based on chord length L), $T_w = T_{o\infty}$ with constant viscosity coefficient (note: in post-processing c_f and c_h, the Sutherland formula was used).

A constant time step $\Delta\tau = 0.1$ is used to solve the full Navier-Stokes equations (1) on the 97 x 33 C-mesh (Fig. 2a). The convergence histories of the upwind and central methods are similar (Figs. 2e, f). The CPU times for 400 time levels on the IBM 3090 computer are 449 and 558 seconds for the central and upwind schemes, respectively.

The pressure increase downstream of about 10 % chord (Fig. 2b) leads to separation from the upper surface at x/L ≈ 0.40 (Fig. 2d). The flow reattaches at x/L ≈ 0.96. Like pressure and skin friction coefficient, even the sensitive heat flux coefficient is predicted in close agreement by the upwind and central methods (Fig. 2c).

HYPERSONIC RAMP FLOW

Laminar flow over a 15° ramp is calculated at $M_\infty = 14.1$, $Re_\infty = 1.04 \times 10^5$ (based on distance L between leading edge and compression corner), $T_w / T_\infty = 4.11538$, $T_\infty = 72.2222\ K$ on the 145 x 65 H-mesh (Fig. 3a) [9]. The local time-stepping option

$$\Delta\tau = \frac{\Delta\tau_{ref}}{\lambda^{(\xi)} + \lambda^{(\eta)}} \tag{11}$$

with $\Delta\tau_{ref} = 2$ is used to solve the thin-layer Navier-Stokes equations for the steady state. The calculations are initialized by interpolating the converged coarse grid solutions. The L_2-norms of the delta variables after 400 time levels are of the order of 10^{-4}. The CPU times on the IBM 3090 computer are about 1250 and 1640 seconds for the central and upwind schemes, respectively. Note that the coefficient in the entropy parameter (9) for the upwind scheme was increased to $\tilde{\delta} = 1/8$ to avoid an overshoot in the Mach number near the leading edge.

The pressure and skin friction coefficients and Stanton number predicted by the upwind and central schemes are in satisfactory agreement with each other, with the results of [16] obtained on a 90 x 30 mesh using the explicit MacCormack scheme, and with measurements [17] (Figs. 3b, c, d). The deviations reflect the sensitivity of the shock boundary layer interaction. The Mach number contours (Figs. 3e, f) indicate the weak leading edge shock and the induced shock, which is captured with fewer grid points by the upwind scheme than by the central method. The overshoot in the maximum Mach number is 0.037 for the upwind scheme, but 0.606 for the central method.

CONCLUSIONS

The upwind method proved in general to be more robust than the central scheme. However, for hypersonic ramp flow, the entropy coefficient in the upwind method had to be increased.

On a coarse mesh, the upwind method predicts the skin friction coefficient for subsonic flat plate flow too low and not as accurately as the central scheme. But the upwind method determines the velocity components near the boundary layer edge and the temperature profile more precisely than the central scheme. On fine meshes, both methods yield almost identical results for subsonic flat plate flow and transonic airfoil flow. For hypersonic ramp flow, the shock resolution with the upwind method is better than with the central scheme.

Using approximate factorization, the upwind method takes about 30 % more CPU time than the central scheme and converges slower. However, the upwind method has the potential of guaranteeing diagonal dominance, and thus allows relaxation methods to be employed efficiently.

REFERENCES

[1] BEAM, R.M., WARMING, R.F.: "An Implicit Factored Scheme for the Compressible Navier-Stokes Equations", AIAA J., 16 (1978), pp. 393-402.

[2] PULLIAM, T.H., STEGER, J.L.: "Recent Improvements in Efficiency, Accuracy, and Convergence for Implicit Approximate Factorization Algorithms", AIAA Paper 85-0360 (1985).

[3] ROE, P.L.: "Characteristic-Based Schemes for the Euler Equations", Ann. Rev. Fluid Mech., 18 (1986), pp. 337-365.

[4] VAN LEER, B., THOMAS, J.L., ROE, P.L., NEWSOME, R.W.: "A Comparison of Numerical Flux Formulas for the Euler and Navier-Stokes Equations", AIAA Paper 87-1104-CP (1987).

[5] HÄNEL, D., SCHWANE, R.: "An Implicit Flux-Vector Splitting Scheme for the Computation of Viscous Hypersonic Flow", AIAA Paper 89-0274 (1989).

[6] YOON, S., KWAK, D.: "Artificial Dissipation Models for Hypersonic External Flow", AIAA Paper 88-3708 (1988).

[7] YEE, H.C.: "Upwind and Symmetric Shock-Capturing Schemes", NASA TM 89464 (1987).

[8] MÜLLER, B.: "Navier-Stokes Solution for Laminar Transonic Flow over a NACA 0012 Airfoil", FFA Report 140, 1986.

[9] MÜLLER, B.: "Calculation of Laminar Supersonic Flows over Ramps and Cylinder Flares", DFVLR IB 221-88 A 14 (1988).

[10] PULLIAM, T.H.: "Artificial Dissipation Models for the Euler Equations", AIAA J., 24 (1986), pp. 1931-1940.

[11] MÜLLER, B., RIEDELBAUCH, S., RUES, D.: "Hypersonic Flow Simulation for Blunt Bodies at Incidence", to appear in: Notes on Numerical Fluid Mechanics, Vieweg, Braunschweig, 1989.

[12] YEE, H.C., HARTEN, A.: "Implicit TVD Schemes for Hyperbolic Conservation Laws in Curvilinear Coordinates", AIAA J., 25 (1987), pp. 266-274.

[13] HARTEN, A.: "High Resolution Schemes for Hyperbolic Conservation Laws", J. Comp. Physics, 49 (1983), pp. 357-393.

[14] YEE, H.C., KLOPFER, G.H., MONTAGNE, J.-L.: "High-Resolution Shock-Capturing Schemes for Inviscid and Viscous Hypersonic Flow", NASA TM 100097 (1988).

[15] MÜLLER, B.: "Simple Improvements of an Upwind TVD Scheme for Hypersonic Flow", to be published in 1989.

[16] HUNG, C.M., MACCORMACK, R.W.: "Numerical Solutions of Supersonic and Hypersonic Laminar Compression Corner Flows", AIAA J., 14 (1976), pp. 475-481.

[17] HOLDEN, M.S., MOSELLE, J.R.: "Theoretical and Experimental Studies of the Shock Wave-Boundary Layer Interaction on Compression Surfaces in Hypersonic Flow", CALSPAN Rept. AF-2410-A-1 (1969).

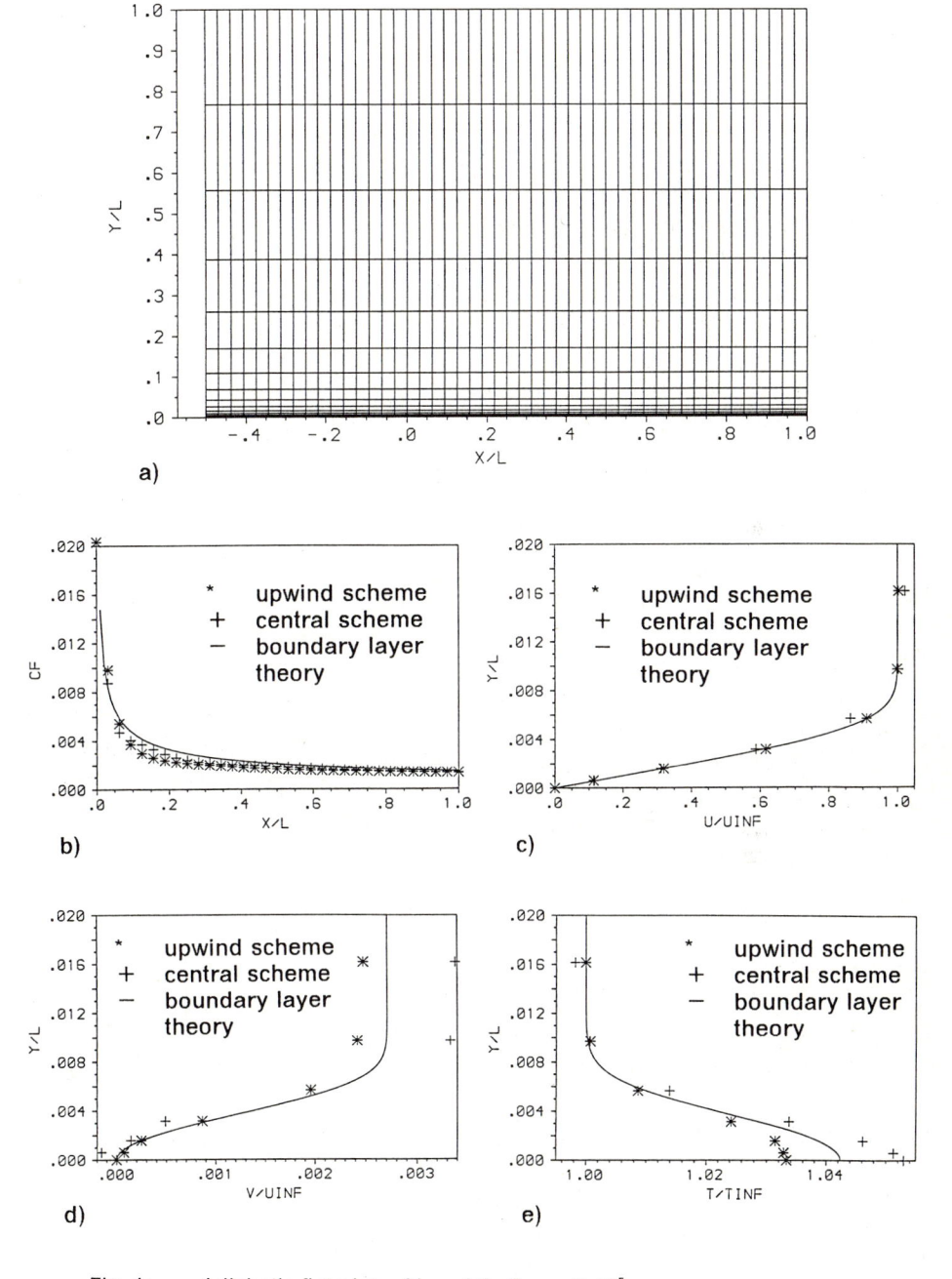

Fig. 1: Adiabatic flat plate, $M_\infty = 0.5$, $Re_\infty = 2x10^5$:
a) 49 x 17 H-mesh,
b) skin friction coefficient, c) u-velocity profile at x/L = 0.5,
d) v-velocity profile at x/L = 0.5, e) temperature at x/L = 0.5.

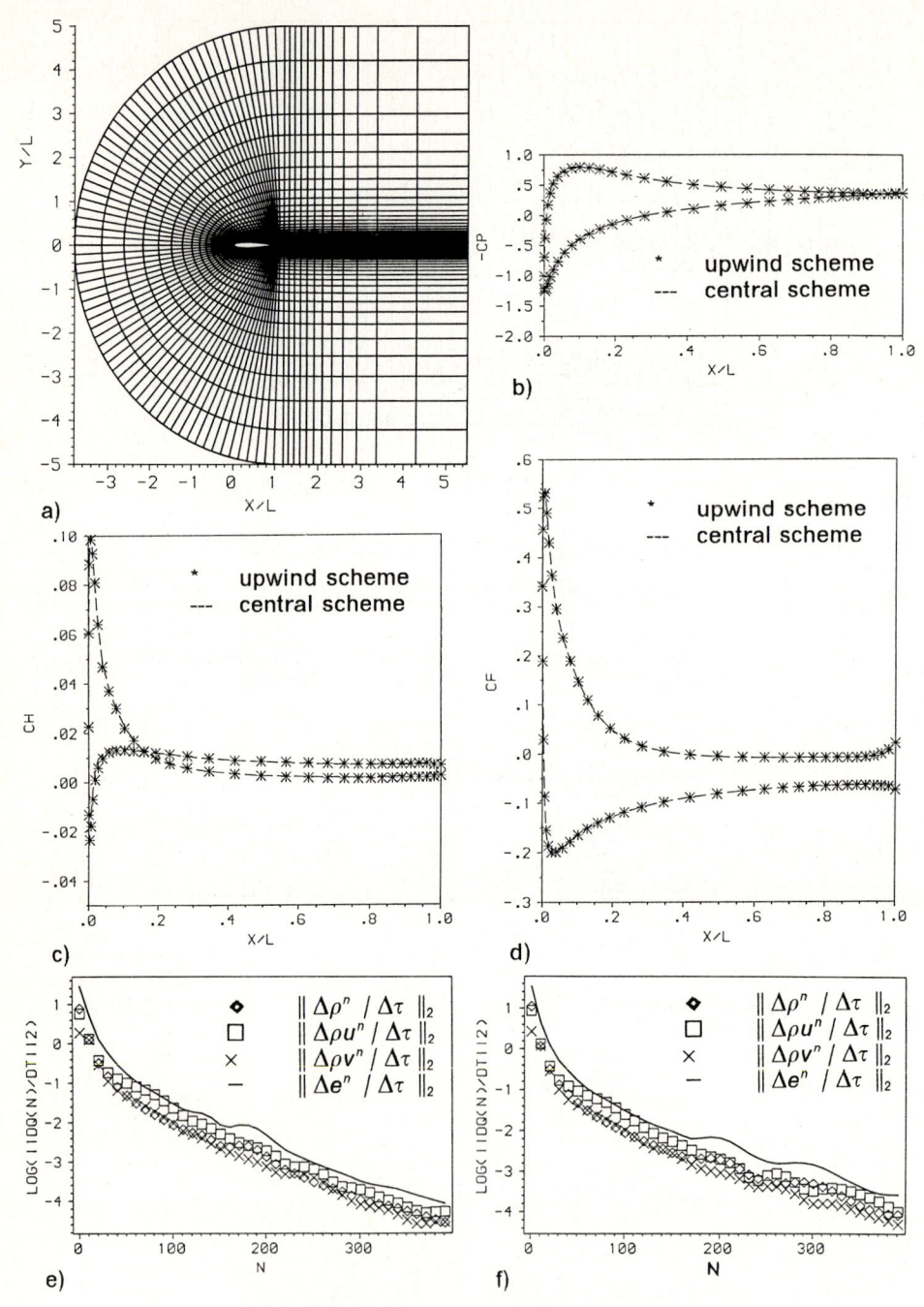

Fig. 2: NACA 0012 airfoil, $M_\infty = 0.8$, $Re_\infty = 500$, $T_w = T_{o\infty}$:
a) 97 x 33 C-mesh, b) pressure coefficient,
c) heat flux coefficient, d) skin friction coefficient,
e) convergence history f) convergence history
for central scheme, for upwind scheme.

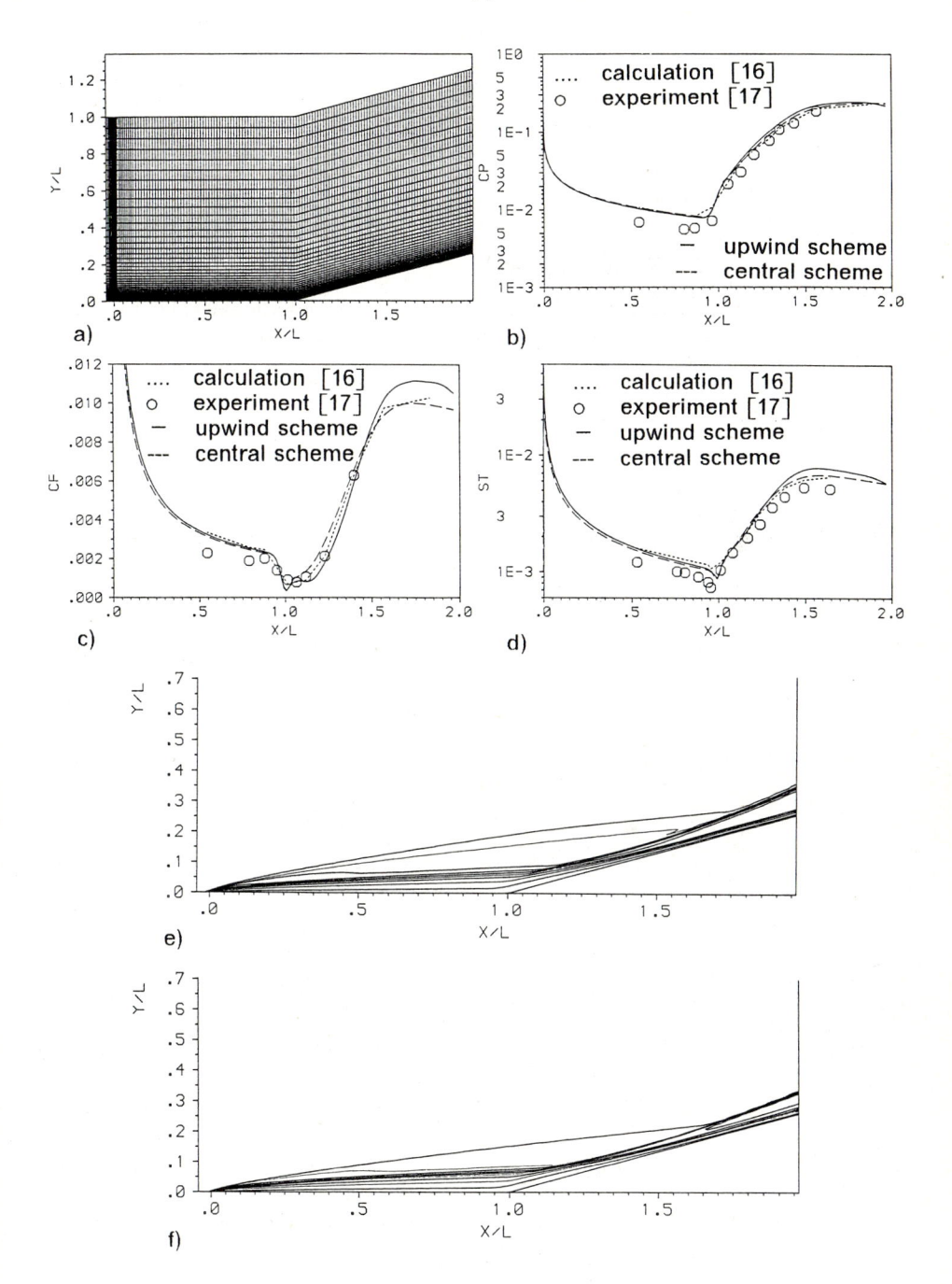

Fig. 3: 15 ° ramp, $M_\infty = 14.1$, $Re_\infty = 1.04 \times 10^5$, $T_w/T_\infty = 4.111538$, $T_\infty = 72.2222\,K$:
a) 145 x 65 H-mesh, b) pressure coefficient,
c) skin friction coefficient, d) Stanton number,
e) Mach number contours with central scheme ($\Delta M = 1.47$, $M_{max} = 14.7$),
f) Mach number contours with upwind scheme ($\Delta M = 1.41$, $M_{max} = 14.1$).

A COMPARISON OF FINITE-DIFFERENCE APPROXIMATIONS FOR THE STREAM FUNCTION FORMULATION OF THE INCOMPRESSIBLE NAVIER-STOKES EQUATIONS

E. Rieger, H. Schütz, D. Wolter, F. Thiele
Technische Universität Berlin,
Hermann-Föttinger-Institut für Thermo- und Fluiddynamik
Straße des 17. Juni 135, D-1000 Berlin 12

ABSTRACT

For the two-dimensional, incompressible Navier-Stokes equations the pure stream function formulation leads to one nonlinear fourth-order differential equation. With Newton linearization the stream function equation can be approximated either by Lagrangian or by Hermitian formulas. The system of finite difference equations is solved directly, taking into account the special structure of the matrix. The main features of the finite difference approximations are discussed with respect to accuracy, stability and computation time. Results are shown and compared with other authors for the steady flow around a circular cylinder.

INTRODUCTION

In order to solve the Navier-Stokes equations one has to make a choice between different formulations. The most frequently used is the vorticity /stream function formulation, in which a set of two second-order partial differential equations for the vorticity and the stream function has to be solved (Examples are given in [1, 2]). Since the vorticity is introduced as an unknown, the vorticity at the boundaries must be specified; usually this is found by iteration.

Expressing the vorticity in terms of the stream function, one nonlinear equation of fourth order is obtained. This is called the stream function formulation and the system of equations resulting from the finite difference approximation can be solved directly. As derivatives up to fourth order appear, the choice of the finite difference formulas becomes important. The method proposed by Schreiber & Keller [3] applies standard 13-point finite difference formulas. They give result for the flow in a driven cavity up to Re = 10,000. Schütz & Thiele [4] use Hermitian polynominals for the approximation of the stream function to calculate the flow around an airfoil at Re = 1,000. The aim of this paper is to compare these different finite difference methods with respect to accuracy, computation time, and error analysis. The steady flow around a circular cylinder at various Reynolds numbers is used as a test case.

GOVERNING EQUATIONS AND BOUNDARY CONDITIONS

Defining the stream function as

$$\mathbf{v} = \nabla \times \mathbf{\Psi} \tag{1}$$

the continuity equation is identically fulfilled and the stream function formulation of the Navier-Stokes equations describing a two-dimensional incompressible steady flow reads

$$\nabla \cdot \left[\frac{1}{2} (\nabla \times \mathbf{\Psi}) \cdot \nabla - \frac{1}{Re} \Delta \right] (\nabla \mathbf{\Psi} - \nabla^T \mathbf{\Psi}) = 0 \; ; \quad \mathbf{\Psi} = (0,0,\psi) \; . \tag{2}$$

The Reynolds number is based on the velocity at infinity and the diameter of the cylinder, $Re = u_\infty d/\nu$. A transformation maps the annular physical flow field (x,y) onto a rectangular computation domain (ξ,η):

$$x = e^\eta \sin(\xi) \; ; \quad y = e^\eta \cos(\xi) \; . \tag{3}$$

The boundary conditions at the cylinder wall are fixed by the no-slip condition and by setting the stream function to zero:

$$\psi_\xi = \psi = 0. \tag{4}$$

An O-type mesh is employed and therefore the periodic boundary condition is applied at the cutting line in front of the cylinder. For the far-field boundary two different types of boundary conditions are implemented. At the far field boundary upstream of the cylinder potential flow is assumed:

$$\psi = (e^\eta - e^{-\eta}) \sin(\xi) \; ; \quad \frac{\partial \psi}{\partial \eta} = (e^\eta + e^{-\eta}) \sin(\xi) \tag{5}$$

while at the far field downstream of the cylinder the velocity and vorticity fields are convected with the velocity at infinity:

$$u_\infty \nabla \Delta \psi = 0 \; ; \quad (\mathbf{u}_\infty \cdot \nabla) \mathbf{v} = 0 \; , \tag{6}$$

FINITE DIFFERENCE METHODS

First the stream function Eq. (2) is linearized according to Newton:

$$\psi^{n+1} = \psi^n + d^n \; , \tag{7}$$

The superscript n indicates the iteration level and d is the corrector function of the stream function ψ. The following describes the Lagrangian- and the Hermitian finite difference methods.

a) Lagrangian Method

Following the suggestion from Schreiber & Keller [3] the stream function is approximated by Lagrangian formulas, resulting in the standard finite difference formulas. The finite difference formula for the fourth derivative of the corrector function, for example, reads

$$d^1_{,\xi\xi\xi\xi} = \frac{1}{\Delta\xi^4}\left[d^{i-2} - 4d^{i-1} + 6d^i - 4d^{i+1} + d^{i+2}\right] - \frac{1}{6}\Delta\xi^2\, d_{,VI}\,. \tag{8}$$

where the superscript i refers to the location in space. The collocation of Eq. (2) leads to a 13-point molecule, shown in Fig. 1.

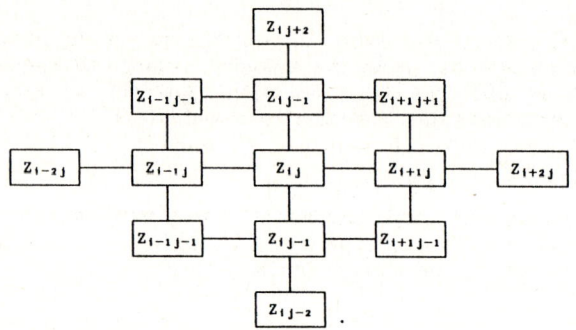

Fig. 1   13-point finite-difference molecule

The resulting truncation errors are of fourth order for the first and second derivatives and of second order for the third and fourth derivatives. At the boundaries Eq. (2) is expressed by nonsymmetric formulas, such as

$$d^1_{,\xi\xi\xi} = \frac{1}{2\Delta\xi^3}\left[-3d^{i-1} + 10d^i - 12d^{i+1} + 6d^{i+2} - d^{i+3}\right] + \frac{1}{4}\Delta\xi^2\, d_{,V} \tag{9}$$

for the third derivative. The finite difference form of Eq. (2) is the well known 13-diagonal N∗M system of equations, where N indicates the number of points in ξ-direction and M the number of points in η-direction. The system is solved applying the Gaussian algorithm with special regard to the band structure [5]. To overcome the nonlinearity the matrix is updated at each iteration.

b) Hermitian Method

The main difference from the method described above lies in the introduction as unknowns of the first derivatives $d_{,\eta}$ and $d_{,\xi}$ of the corrector function d. This permits the application of the so-called compact finite difference formulas, such as

$$d^1_{,\xi\xi\xi\xi} = -\frac{12}{\Delta\xi^4}\left[d^{i+1} - 2d^i + d^{i-1}\right] + \frac{6}{\Delta\xi^3}\left[d^{i+1}_{,\xi} - d^{i-1}_{,\xi}\right] - \frac{\Delta\xi^2}{15}\,d_{,VI} \tag{10}$$

for the fourth derivative. This leads to a nine-point molecule with three unknowns each grid point:

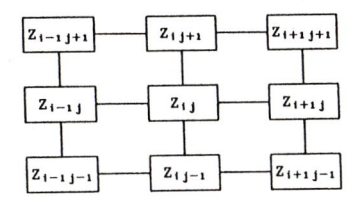

Fig. 2   9-point finite difference molecule

Now the boundary conditions for ψ, $\psi_{,\eta}$ and $\psi_{,\xi}$ can be incorporated directly into the matrix system. Furthermore, the whole flow field can be approximated by symmetric finite difference formulas, even next to the boundaries. The first derivatives are unknowns and are therefore exact, while the second and third derivatives contain a truncation error of fourth order and the fourth derivative of second order.

For $d_{,\xi}$ and $d_{,\eta}$ we apply "Mehrstellen" formulas in each direction:

$$d_{,\xi}^{i-1} + 4d_{,\xi}^{i} + d_{,\xi}^{i+1} + \frac{1}{\Delta\xi}(d^{i-1} - d^{i+1}) = O(\Delta\xi^4)$$

$$d_{,\eta}^{j-1} + 4d_{,\eta}^{j} + d_{,\eta}^{j+1} + \frac{1}{\Delta\eta}(d^{j-1} - d^{j+1}) = O(\Delta\eta^4) .$$

(11)

As shown by Wagner [5] the resulting block-tridiagonal system of equations reads

Fig. 3   Block-tridiagonal matrix system

The first and last line of each block contain the coefficients for the "Mehrstellen" formulas in ξ- and η-direction. The linearized finite difference form

The calculation for the first row of separation lengths shown in Table 1 are based on a grid with 51·51 points and a ratio of the outer to the inner grid radius r_∞/r = 54.6. Unfortunately, the calculation for Re = 100 did not converge, so a smaller grid was used with 98·83 points and r_∞/r = 148.4. The number of iterations required for convergence was 10-11 with 32.4 CPU-seconds per iteration step for the Lagrangian method and 3 Newton / 11 Chord steps with 162.3 CPU-s / 1.93 CPU-s for the Hermitian method (second row of Table 1). The Hermitian results are in better agreement with the measured separation lengths (Coutanceau & Bouard [8]) than Lagrangian results. This could be anticipated from the streamline and vorticity patterns for the calculations on the fine grid (Fig. 3), which shows wiggles increasing with the Reynolds number for the Lagrangian method compared with much smoother contours for the Hermitian.

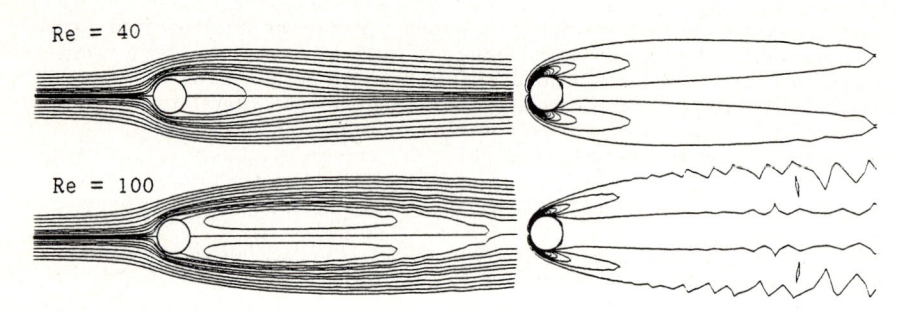

Re = 40

Re = 100

a) Lagrangian method

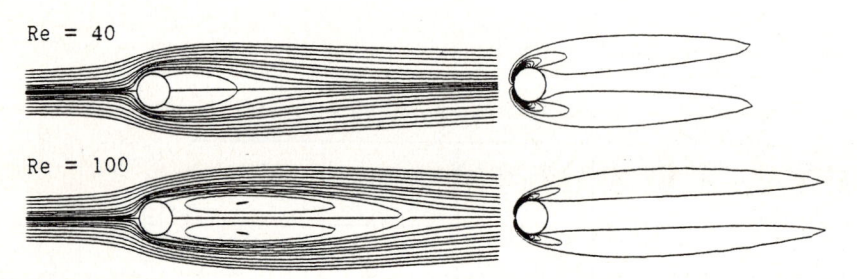

Re = 40

Re = 100

b) Hermitian method

Fig. 4: Streamline and vorticity patterns on the fine grid

A nonsymmetric condensed grid (Fig. 5) was produced in order to study the influence of the transformation. Fig. 6 shows in the Lagrangian method the separating streamline is aligned with the grid lines of highest density. The Hermitian solution does not seem to be affected by the transformation, which becomes important in cases with unknown preferential flow direction, for instance the unsteady flow around an airfoil.

Even if the grid is condensed symmetrically, distorsion of the grid in the circumferential direction influences the Lagrangian results. This is demonstrated in Fig. 7, where the contours of vorticity for Re = 0 are plotted. The calculations are done on the grid containing 98·83 points as used before. While the contours remain axisymmetric for the Hermitian method they are of Eq. (2) occurs in the middle line of each block. The system of equations is solved with an efficient L-U decomposition adapted to the block-tridiagonal matrix structure. For the nonlinearity the Newton-Chord iteration technique is applied, where the time-consuming matrix decomposition is done only for a Newton step. For a Chord step the decomposed matrix from the previous Newton step is taken and only the righthand side is updated.

To summarize, the Hermitian method differs from the Lagrangian method by the introduction of the first derivatives as unknowns; these new variables have the physical interpretation of transformed velocities. This simplifies the calculation of further important parameters, for example the shearstress coefficient, where one numerical derivative fewer has to be computed than for the Lagrangian method. Neumann boundary conditions can be expressed exactly in the matrix, which is advantageous for boundaries with an imposed velocity, like the no-slip condition. Due to the smaller finite difference molecule, no unsymmetric finite difference formulas are necessary; this means the approximation of the wall region, where high gradients occur, is fully consistent with the inner region of the flow field.

An analysis of the truncation error (not given here) shows that the convective term error, which grows with increasing Reynolds number, is two orders of magnitude smaller than for the Lagrangian method. Furthermore, the truncation error is more stiff because the stream function itself and the first derivatives are fixed. Finally, the difference molecule occupies only the point of collocation and its nearest neighbors. The influence of these differences are discussed in the following section.

NUMERICAL COMPARISON

All calculations were carried out on a CRAY X-MP/24. The results for the separation lengths are compared with values found in the literature in Table 1.

Table 1   List of Separation Lengths

Author Re =	20	40	100

Present 51*51 r_∞/r = 54.6
    Lagrange
    Hermite
	
	0.527
	0.664
		
		1.45
		1.73
			
			n.c.
			8.02

Present 99*83 r_∞/r = 148.4
    Lagrange
    Hermite
	
	0.753
	0.813
		
		1.93
		2.08
			
			9.65
			7.09

Coutanceau & Bouard (M) [8]	0.87	2.05	6.85
Takami & Keller from [7]	0.93	2.33	-
Nieuwstadt & Keller from [7]	0.89	2.17	-
Ta & Phuoc from [7]	0.91	2.24	6.35

(M) Measurement;   n.c. no convergence

distorted for the Lagrangian method.

Fig. 5: Non-symmetric condensed grid with 55·53 points

Fig. 6: Streamline pattern Re = 40

a) Lagrangian method

b) Hermitian method

a) Lagrangian method

b) Hermitian Method

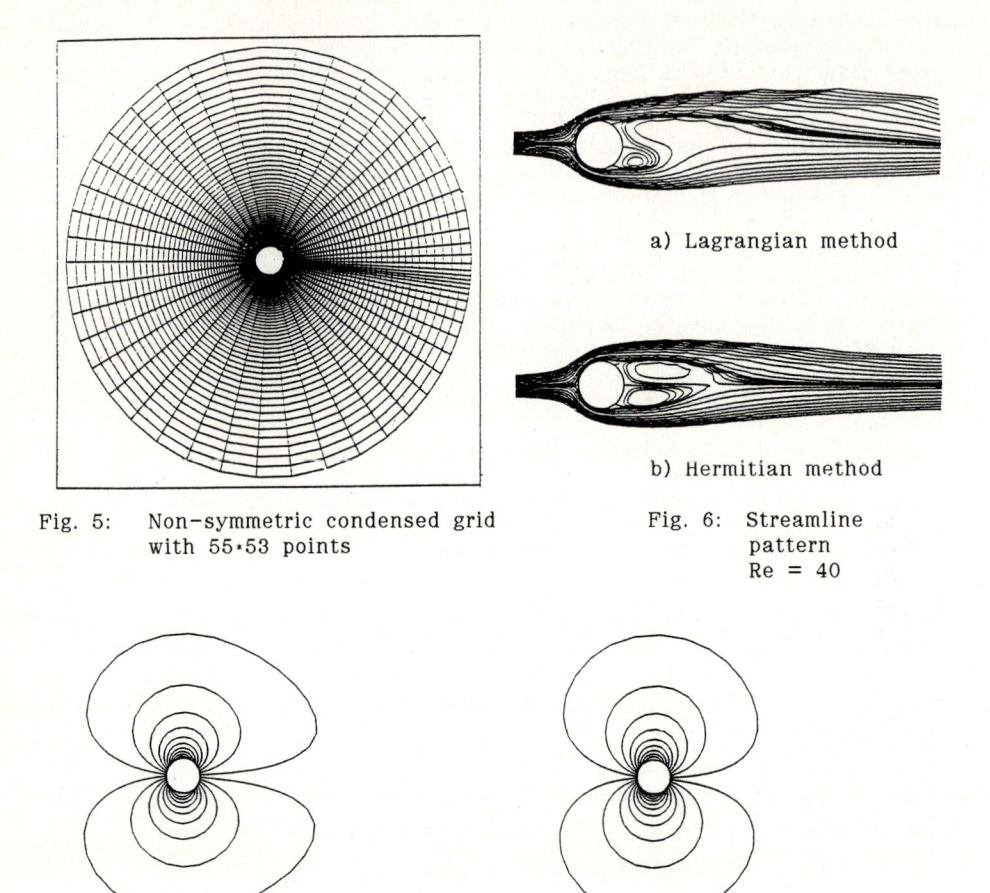

Fig. 7: Isolines of vorticity for Re = 0

Finally, in Fig. 8, the error properties of the two methods are investigated. In order to force pertubations a grid with a very small outflow domain, located here at r_∞/r = 14.9, and 51·51 points was used. The vorticity in the computational domain is shown as a 3-d plot for Re = 100. Because the assumed potential flow is not really valid at the farfield boundary the switch of the far-field boundary condition from Eq. (5) to Eq. (6) produces pertubations. These pertubations are reflected into the inner flow field in the Lagrangian case. The Hermitian vorticity field looks smoother which may result from the stiffer error function as discussed earlier. This is advantageous for unsteady flow problems where the solution is requested to be free of effects from the non-physical outflow boundary condition.

a) Langrangian method   b) Hermitian method

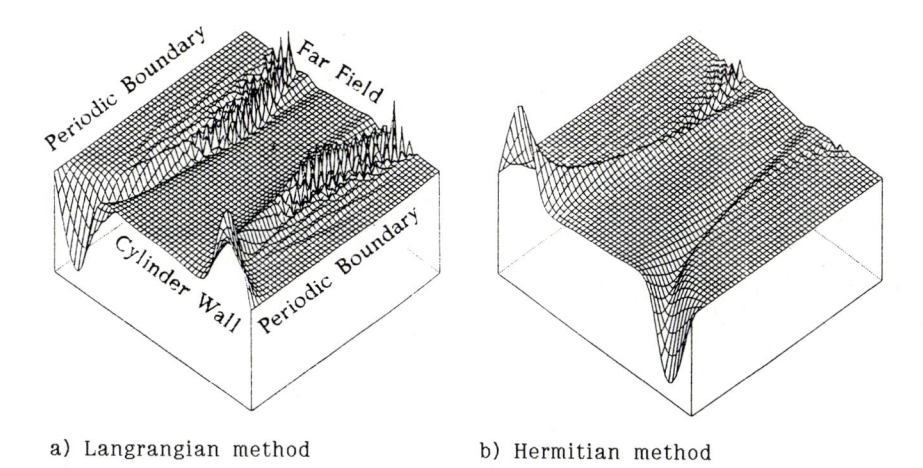

Fig. 8: 3-d plot of the vorticity for Re = 100

CONCLUSIONS

Two different finite difference formulations of the incompressible Navier-Stokes equations in the stream function formulation have been compared. The introduction of the velocities as additional unknowns increases the CPU time consumption for the Hermitian method. In return the Hermitian method gives more stable results, even for a smaller number of grid points. Because of the smaller difference molecule needed for the Hermitian approximation of the Navier-Stokes equations, the influence of the transformation is less than for the Lagrangian method. Furthermore pertubations, such as produced by the outflow boundary conditions, remain local for the Hermitian method.

REFERENCES

[1] BRILEY, W.R.: "A Numerical Study of Laminar Separation Bubbles using the Navier-Stokes Equations", J. Fluid Mech., 47 (1971), pp. 58-68.

[2] LECOINTE, Y., PIQUET, J.: "On the Use of Several Compact Methods for the Study of Unsteady Incompressible Viscous Flow Around a Circular Cylinder", Comp. & Fluids 12 (1984) pp. 255-260.

[3] SCHREIBER, R., KELLER, H.B.: "Driven Cavity Flows by Efficient Numerical Technics", J. Comp. Physics, 49 (1983) pp. 310-333.

[4] SCHÜTZ, H., THIELE, F.: "Unsteady 2-dimensional flow around bodies using the Navier-Stokes equations". In: Taylor, C., Habashi, W.C., Hafez, M.M. (eds.): Numerical Methods in Laminar and Turbulent Flows, Procs. 5th Int. Conf. Montreal (1987).

[5] SCHRAUF, G.: "A Gauss Algorithm to Solve Systems with Large, Banded Matrices Using Random Access Disk Storage", ACM Transactions on Math. Software $\underline{3}$ (1988) pp. 257-260.

[6] WAGNER, H.: "Ein Differenzenverfahren für die Navier-Stokes Gleichungen in mehrfach zusammenhängenden Gebieten", Doctoral Thesis, Technische Universität Berlin (1984).

[7] FORNBERG, B.: "A Numerical Study of Steady Viscous Flow Past a Circular Cylinder", J. Fluid Mech., $\underline{98}$ (1980) pp. 819-855.

[8] BOUARD, R., COUTANCEAU, M.: "Experimental Determination of the Main Features of the Viscous Flow in the Wake of a Circular Cylinder in Uniform Translation, Part 1 - Steady Flow", J. Fluid Mech. $\underline{79}$ (1977) pp. 231-272.

NSFLEX – AN IMPLICIT RELAXATION METHOD FOR THE NAVIER-STOKES EQUATIONS FOR A WIDE RANGE OF MACH NUMBERS

M.A. SCHMATZ
MESSERSCHMITT-BÖLKOW-BLOHM GmbH,
FE122, Postfach 801160
D – 8000 München 80, FRG

SUMMARY

Discussed is the well-proven NSFLEX method, which was applied to various sub- and transonic flow cases in the past. Here the extension to hypersonic flows is discribed. Applications are given for simple two-dimensional as well as for complex three-dimensional configurations. Real gas effects are incorporated in the solution procedure. For turbulent flows the Reynolds-averaged Navier-Stokes equations are employed using an algebraic turbulence model. To evaluate the inviscid fluxes a Riemann problem is solved at the finite-volume faces. A third-order accurate local characteristic flux extrapolation scheme (MUSCL type flux difference splitting) is utilized, using van Albada sensors to detect non-monotonous behaviour of the flow variables, where the scheme degrades to a first-order one. At very strong shocks, which occur at hypersonic flow conditions, a hybrid Steger-Warming (flux vector splitting) local characteristic flux is used to avoid negative pressures in the transient phase. For the present application the Steger-Warming flux is modified to overcome some disadvantages of the original one. Up to third-order accuracy is employed to reduce the inherent numerical viscosity of the inviscid flux discretisation. The viscous fluxes are constructed with central differences at each cell face. The unfactored implicit equations are solved in time-dependent form by a point Gauss-Seidel relaxation technique with red-black strategy. Thereby the code is perfectly vectorized. Because it is a finite-volume scheme complex geometries can be handled. Highly accurate results are given in the present paper for very low and very high Mach number applications.

INTRODUCTION

Various applications of the **NSFLEX** method (**N**avier-**S**tokes solver using characteristic **fl**ux **ex**trapolation) were performed in the past for sub-, trans- and slightly supersonic two- and three-dimensional visous flow cases [1,2]. A typical two-dimensional transonic verification example is reproduced in Fig. 1 and Fig. 2 [2]. The freestream Mach number is $M_\infty=0.73$, the Reynolds number Re=6,000,000 and transition is fixed on the upper and on the lower side, respectively, at 7 per cent of the chord length. The angle of attack is $\alpha=2.1°$. The C-type grid consists of 248 times 49 cells. In Fig. 1 the iso-Mach lines demonstrate the smoothness of the solution. In Fig. 2 the computed pressure distribution compares very well with the experimental one. Al-

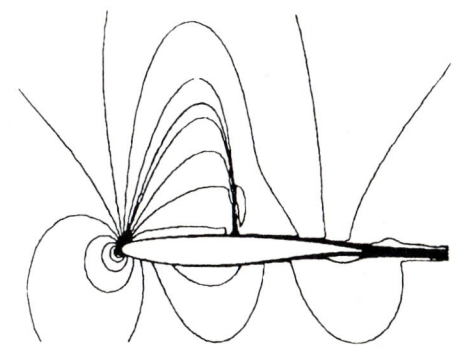

Fig. 1   Iso-Mach lines ($M_\infty=0.73$, Re=6,000,000, $\alpha=2.1°$)

so the computed aerodynamic forces are in good agreement with the experiment [2]. Problems still exist for highly separated flow cases due to the turbulence model [2].

The code also was extensively verified and applied in a zonal Euler-, boundary-layer-, Navier-Stokes version [3,4].

First tests to calculate hypersonic flow cases, where the Mach number is higher than about 5 and where sharp discontinuities exist in the flow variables, failed with the present flux difference splitting method. Therefore the so-called Steger-Warming flux vector splitting scheme was added to the original method as a remedy where stability problems occur.

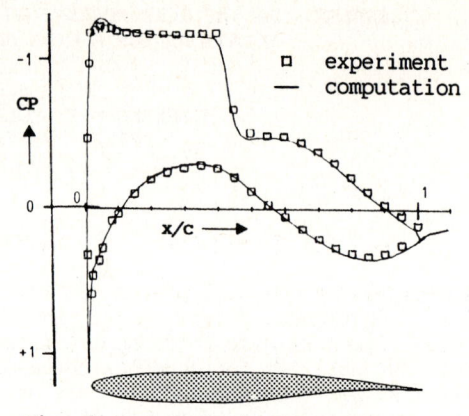

Fig. 2 Pressure distribution (M_∞=0.73, Re=6,000,000, α=2.1°)

GOVERNING EQUATIONS

The governing equations are the time dependent compressible Navier-Stokes equations in conservation law form. To simulate turbulent flows the Reynolds-averaged equations are closed with an algebraic turbulence model

$$U_t + E_\xi + F_\eta + G_\zeta = 0, \quad \text{where:} \quad U = J(\rho, \rho u, \rho v, \rho w, e)^T. \quad (1)$$

Body-fitted arbitrary coordinates ξ, η, ζ with cartesian velocity components u,v,w are used. The fluxes normal to ξ=const., η=const., ζ=const. finite-volume faces are:

$$E = J(\tilde{E}\xi_x + \tilde{F}\xi_y + \tilde{G}\xi_z), \qquad F = J(\tilde{E}\eta_x + \tilde{F}\eta_y + \tilde{G}\eta_z), \quad (2)$$

$$G = J(\tilde{E}\zeta_x + \tilde{F}\zeta_y + \tilde{G}\zeta_z),$$

with the Cartesian fluxes:

$$\tilde{E} = \begin{bmatrix} \rho u \\ \rho u^2 - \sigma_{xx} \\ \rho uv - \sigma_{xy} \\ \rho uw - \sigma_{xz} \\ (e-\sigma_{xx})u - \sigma_{xy}v - \sigma_{xz}w + q_x \end{bmatrix}, \quad \tilde{F} = \begin{bmatrix} \rho v \\ \rho vu - \sigma_{yx} \\ \rho v^2 - \sigma_{yy} \\ \rho vw - \sigma_{yz} \\ (e-\sigma_{yy})v - \sigma_{yx}u - \sigma_{yz}w + q_y \end{bmatrix}, \quad (3)$$

$$\tilde{G} = \begin{bmatrix} \rho w \\ \rho wu - \sigma_{zx} \\ \rho wv - \sigma_{zy} \\ \rho w^2 - \sigma_{zz} \\ (e-\sigma_{zz})w - \sigma_{zx}u - \sigma_{zy}v + q_z \end{bmatrix}.$$

The stress tensor is:

$$\sigma_{xx} = -p - \frac{2}{3}\mu(-2u_x+v_y+w_z), \qquad \sigma_{yy} = -p - \frac{2}{3}\mu(u_x-2v_y+w_z), \qquad (4)$$

$$\sigma_{zz} = -p - \frac{2}{3}\mu(u_x+v_y-2w_z),$$

$$\sigma_{xy} = \sigma_{yx} = \mu(u_y+v_x), \qquad \sigma_{xz} = \sigma_{zx} = \mu(w_x+u_z),$$

$$\sigma_{yz} = \sigma_{zy} = \mu(v_z+w_y),$$

and the heat flux vector:

$$q_x = -kT_x, \qquad q_y = -kT_y, \qquad q_z = -kT_z. \qquad (5)$$

ρ, p, T, μ, k are density, pressure, temperature, coefficient of viscosity and heat conductivity coefficient. The indices $()_\xi, ()_\eta, ()_\zeta, ()_x, ()_y, ()_z$ denote partial derivatives with respect to ξ, η, ζ or x, y, z except for the stress tensor σ and the heat flux vector q. Effective transport coefficients are introduced with the Boussinesq approximation. The metric factors are given for example in [1].

TIME INTEGRATION

To reach the steady state solution an implicit relaxation procedure for the unfactored equations is used which allows high CFL numbers [1,2]. With the first order in time discretized implicit form of Eq. (1)

$$(U^{n+1} - U^n)/\Delta t + E_\xi^{n+1} + F_\eta^{n+1} + G_\zeta^{n+1} = 0 \qquad (6)$$

a Newton method can easily be constructed for U^{n+1} by linearizing the fluxes of Eq. (6) about the known time level n:

$$E^{n+1} = E^n + A^n \Delta U, \qquad F^{n+1} = F^n + B^n \Delta U, \qquad G^{n+1} = G^n + C^n \Delta U, \qquad (7)$$

leading to

$$\Delta U/\Delta t + (A^n \Delta U)_\xi + (B^n \Delta U)_\eta + (C^n \Delta U)_\zeta = -(E_\xi + F_\eta + G_\zeta)^n = \text{RHS}. \qquad (8)$$

Therein A, B, C are the Jacobians of the flux vectors E, F, G :

$$A = \partial E/\partial U, \quad B = \partial F/\partial U, \quad C = \partial G/\partial U. \qquad (9)$$

ΔU is the time variation of the solution and therefore the update is

$$U^{n+1} = U^n + \Delta U. \qquad (10)$$

To construct a relaxation method it is necessary to have a diagonal dominant system of equations which is achieved with upwind differencing of the inviscid flux vectors. Special care on using a **true** representation of the fluxes in the left-hand-side **Jacobians** allows the use of high CFL numbers [1,2,5]. For the fluxes used in the present scheme this property is inherent. With the divergence of the fluxes on the right-hand-side (RHS), Eq. (8) has to be solved approximately at every time step. A point Gauss-Seidel technique is used, where it is not necessary to balance the equation perfectly. The terms $(A^n \Delta U)_\xi$, $(B^n \Delta U)_\eta$, $(C^n \Delta U)_\zeta$ on the LHS are discretized at i,j,k up to second order in space [1], for example $(A^n \Delta U)_\xi$ along j,k=const

$$(A^n \Delta U)_\xi = (A_{inv}^n \Delta U)_{i+1/2} - (A_{inv}^n \Delta U)_{i-1/2} + (A_{vis}^n)_i (\Delta U_{i+1} - 2\Delta U_i + \Delta U_{i-1}), \quad (11)$$

where:

$$(A_{inv}^n \Delta U)_{i+1/2} = (T\Lambda^+ T^{-1})_{i+1/2} \Delta U^+_{i+1/2} + (T\Lambda^- T^{-1})_{i+1/2} \Delta U^-_{i+1/2}, \quad (12)$$

$$(A_{inv}^n \Delta U)_{i-1/2} = (T\Lambda^+ T^{-1})_{i-1/2} \Delta U^+_{i-1/2} + (T\Lambda^- T^{-1})_{i-1/2} \Delta U^-_{i-1/2},$$

with ΔU vectors extrapolated consistently to the right-hand-side [1]. For the finite-volume face at i+1/2,j,k ΔU is extrapolated up to second order depending on the eigenvalues and the sensor β of the RHS (see the following Chapter). A_{vis} is the thin-layer viscous Jacobian at i,j,k for all directions ξ, η, ζ. Λ is the diagonal matrix of the eigenvalues of the Jacobian A. T and T^{-1} are the matrices which diagonalize A ($\Lambda = T^{-1}AT$). The coefficients of T, Λ, T^{-1} are found from arithmetic means of the conservative variables $\rho, \rho u, \rho v, \rho w, e$. Λ^+ is the diagonal matrix of the positive eigenvalues of A_{inv} and Λ^- the matrix of the negative ones

$$\Lambda^+ = \max(E, \Lambda), \quad \Lambda^- = \min(-E, \Lambda). \quad (13)$$

The term E in Eq. (13) is zero where β is zero and leads to an additional residual smoother at non-monotonous flow regions

$$E = \beta * SMREL * \max(abs(\lambda_4), abs(\lambda_5)). \quad (14)$$

β is zero for smooth flow situations and one for non-monotonous ones. SMREL is typically set to 0.3. With these discretisation formula applied to the three spatial directions Eq. (8) is reordered for point Gauss-Seidel iteration. The iteration count is indicated by an upper index μ

$$DIAG_{i,j,k}^n \Delta U_{i,j,k}^{\mu+1} = -\omega \, RHS_{i,j,k}^n + ODIAG_{i,j,k}. \quad (15)$$

$DIAG_{i,j,k}^n$ is a 5*5 matrix of the sum of the eigenvalue splitted inviscid and the viscous thin-layer Jacobians together with the inverse of time step $I/\Delta t$. ω is an underrelaxation parameter compensating errors of different spatial order of accuracy on RHS and LHS (.3<ω<1). The time step is calculated with the maximum of the eigenvalues of the inviscid Jacobians

$$\Delta t = (\beta * CFLMIN + (1.-\beta) * CFL) / (\max |\lambda_{i,j,k}|). \quad (16)$$

The CFL number is typically 150 to 200 for moderate Mach numbers and reduces to lower ones for hypersonic applications. CFLMIN limits the time step at regions where strong shocks appear. The fluxes $RHS_{i,j,k}^n$ and the matrix $DIAG_{i,j,k}^n$ rest at time level n during μ-iteration. The term ODIAG is given by

$$ODIAG = f(\Delta U_{i+1,j,k}, \Delta U_{i-1,j,k}, \Delta U_{i,j+1,k}, \Delta U_{i,j-1,k}, \quad (17)$$
$$\Delta U_{i,j,k+1}, \Delta U_{i,j,k-1}),$$

where the actual ΔU values of the μ-iteration are taken. Three Gauss-Seidel steps (μ=3) are performed at every time step for the present calculations. Two features of the scheme are important to allow high CFL numbers: True Jacobians of the fluxes on the RHS must be used on the LHS. A pre-conditioning of the matrix $DIAG_{i,j,k}$ is done such that diagonal dominance is forced [1,5]. Locally the system is solved for the non-conservative ΔU* vector and transformed back to a conservative ΔU at every point i,j,k. This transformation does not influence conservation properties of the scheme at all [1,5].

FLUX CALCULATION

For the inviscid flux calculation a linear locally one-dimensional Riemann solver for E,F,G is employed at the finite-volume faces. Since the code should be able to work for a wide range of Mach numbers (.05 < Mach < 100), a hybrid local characteristic (LC) [6] and Steger-Warming type (SW) flux scheme is employed, see also [11]. For example the flux at cell face i+1/2 is found

$$E_{i+1/2} = (E_{LC}(1-a) + a*E_{SW})_{i+1/2} , \qquad (18)$$

with $a=SW*s*(M_r-M_l)^2$, where s is the van Albada sensor [7] for the density, M_r and M_l are the Mach numbers on the left- and on the right-hand-side of the volume face. SW is an input constant to be specified between zero and an arbitrary high number. To calculate the local charcteristic fluxes, the conservative variables on either side of the volume face are extrapolated up to third order in space (MUSCL type extrapolation):

$$U_l = U_i + s/4*((1-k*s)\Delta_- + (1+k*s)\Delta_+)_i , \qquad (19)$$

$$U_r = U_{i+1} - s/4*((1-k*s)\Delta_+ + (1+k*s)\Delta_-)_{i+1} ,$$

where k=1/3 for third order upwind biased extrapolation. s is a van Albada type sensor [7]

$$s=(2\Delta_+\Delta_- + \varepsilon)/((\Delta_+)^2+(\Delta_-)^2+\varepsilon), \text{ with } \Delta_+=U_{i+1}-U_i , \Delta_-=U_i-U_{i-1} . \qquad (20)$$

ε is a small number preventing division by zero ($\varepsilon=5.10^{-5}$). The number β for use on the left-hand-side is β=max(a,s) (a from Eq. (18) and s from Eq. 20)). With the variables U_l and U_r an eigenvalue weighted mean value is found at the volume face

$$U_{i+1/2} = T_{i+1/2} \Sigma_l (\Lambda'^+_l T^-_{i+1/2} U_l + \Lambda'^-_l T^-_{i+1/2} U_r) . \qquad (l=1,5) \qquad (21)$$

With this vector the inviscid fluxes can be calculated directly (flux difference splitting)

$$E_{LC} = E(U_{i+1/2}) . \qquad (22)$$

This scheme guarantees the homogeneous property of the Euler fluxes, a property which simplifies the evaluation of the true Jacobians of the fluxes for use on the left-hand-side [1,5].

Since this local characteristic flux is not diffusive enough to guarantee stability for hypersonic flow cases especially in the transient phase, where the shocks move, at regions of high gradients a Steger-Warming [8] type flux is utilized, locally (flux vector splitting)

$$E_{SWi+1/2} = E_i^+ + E_{i+1}^- = (T\Lambda^+ T^- U)_i + (T\Lambda^- T^- U)_{i+1} . \qquad (23)$$

Here some modifications of the original Steger-Warming fluxes are introduced:

$$\lambda_{1,2,3}^+ = \lambda_3^+ + \lambda_4^+ ,$$
$$\lambda_{1,2,3}^- = \lambda_3^- + \lambda_4^- ,$$
(24)

Energy flux = mass flux * H_{tot} . (25)

Eq. (25) gives a better conservation of the total energy as was also found in [9] for the van Leer flux vector splitting scheme. Eq. (24) avoids the unsteadyness of the mass flux in the original Steger-Warming flux, see Fig. 3.

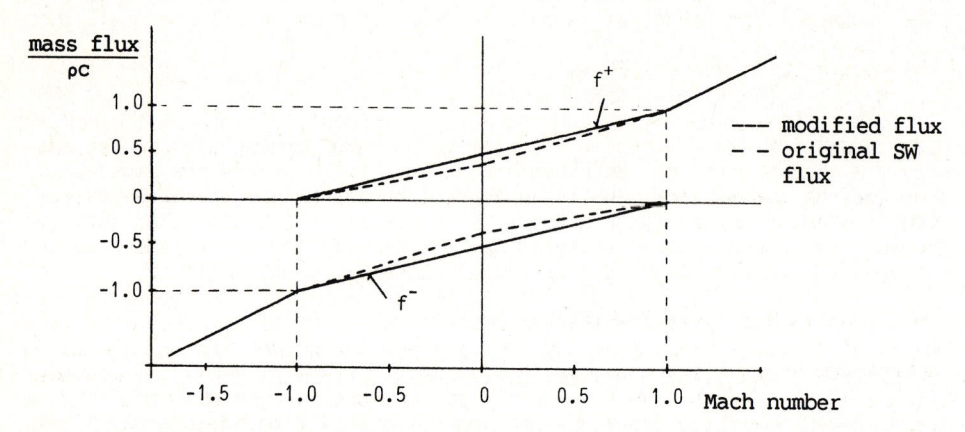

Fig. 3 Variation of original and modified Steger-Warming splitted mass fluxes with Mach number

Diffusive fluxes at the cell faces are calculated with central differences, e.g. for the ξ=const. cell face

$$\phi = \phi_{i+1,j,k} - \phi_{i,j,k} .$$
(26)

ϕ represents the velocity components u,v,w or the temperature T. For the cross derivatives, where ϕ_η and ϕ_ζ are needed, conventional differencing is used:

$$\phi_\eta = (\phi_{i+1/2,j+1,k} - \phi_{i+1/2,j-1,k})/2 ,$$
(27)

$$\phi_\zeta = (\phi_{i+1/2,j,k+1} - \phi_{i+1/2,j,k-1})/2 .$$

$\phi_{i+1/2,j+1,k}$, $\phi_{i+1/2,j-1,k}$, $\phi_{i+1/2,j,k+1}$, $\phi_{i+1/2,j,k-1}$ are found as arithmetic means of neighbouring cell center values.

EQUILIBRIUM REAL GAS INCORPORATION

In the Euler equations the ratio of the specific heats appears only in the energy equation, writen here for simplicity in cartesian coordinates

$$(e_r)_t + (u(p+e_r))_x + (v(p+e_r))_y + (w(p+e_r))_z = 0 ,$$
(28)

with

$$e_r = p/(\gamma_r-1) + \rho q^2/2 \, , \qquad \gamma_r = \gamma_r(p,\rho) \; . \qquad (29)$$

The index r denotes real equilibrium gas. $\gamma_r = \gamma_r(p,\rho)$ is calculated from a thermodynamic subroutine of SRINIVASAN et al. [10]. Following EBERLE [11] we define a new total energy e with a reference ratio of specific heats γ which is the freestream γ

$$e = p/(\gamma-1) + \rho q^2/2 \; . \qquad (30)$$

With:

$$\begin{aligned} e_r &= e + Q \, , \\ Q &= e_r - e = p(1/(\gamma_r-1) - 1/(\gamma-1)) \end{aligned} \qquad (31)$$

the energy equation is

$$(e)_t + (Q)_t + (u(p+e))_x + (v(p+e))_y + (w(p+e))_z = \qquad (32)$$
$$- ((u*Q)_x + (v*Q)_y + (w*Q)_z) \; .$$

Since we are just interested in the steady state solution we can use a steadiness assumption for Q

$$(Q)_t = 0 \; . \qquad (33)$$

This means that the left-hand-side of Eq. (33) is the perfect gas energy equation and the real gas influence is separated on the right-hand-side as a source term. This pseudo-unsteady approach now offers the big advantage that neither the Riemann solver nor the implicit part of the NSFLEX solver has to be changed.

To account for the effects of deviation from perfect gas assumption in the viscous fluxes some more curve fits [10,12] for the temperature and the transport coefficients have to be used

$$T = T(p,\rho) \, , \quad \mu = \mu(p,\rho) \, , \quad k = k(p,\rho) \; . \qquad (34)$$

In the present calculations real gas incorporation slows down the NSFLEX method by a factor of about two due to the fact that the curve fit routines are not vectorized at all. Therefore work is under progress to create new splines which are vectorizable [13], which will replace the curve fits given in [10,12].

APPLICATIONS

First a low Mach number simulation is shown. It is the flow past a flapped airfoil [14] with: freestream Mach number $M_\infty=0.1$, Reynolds number Re=6,000,000, angle of attack $\alpha=13.21°$. Transition is fixed on the upper and on the lower side, respectively, at 1 per cent of the chord length. The algebraic turbulence model of BALDWIN-LOMAX is employed. In Fig. 4 the NSFLEX pressure distribution [15] is compared with the experiment [14]. Good agreement can be observed. Separation occurs at the rear part of the airfoil, see Fig. 5 for the streamlines.

A second two-dimensional case validates the method for hypersonic Mach numbers and shows also the small amount of inherent non-physical viscosity. In Fig. 6 the upper half of the grid for a hyperbola is shown. The mesh is adapted to the solution in the boundary-layer regime and at the bow shock position. Around the hyperbola 240 cells and normal to the body 50 cells are distributed. Note that the calculation was performed with no symmetry condition although the flow is symetrical to check whether the code works symmetrically. The freestream Mach number for this laminar case is $M_\infty=10.$, the Reynolds number per meter Re/m=1,600,000, the angle of attack is $\alpha=0.0°$. The freestream temperature $T_\infty=220K$, adiabatic wall is assumed together with perfect gas. Both the isobars (Fig. 7) and the iso-Mach lines (Fig. 8) demonstrate the smoothness of the solution and the perfect shock capturing feature of the method. In Fig. 9,10,11 the temperature, the skin friction coefficient and the pressure distribution are plottet against the body axis. The stagnation temperature is reproduced very accurately. Computed is a temperature of $T_{stag}=4618.98K$, whereas the analytical one is $T_{stag}=4620K$.

An application to the HERMES configuration is shown in Figs. 12,13, 14,15. The surface grid together with the symmetry plane and an x=const. cut is shown in Fig. 12. The flow charcteristics are: freestream Mach number $M_\infty=10$, Reynolds number referenced to the total body length Re=130,000 and 13,000,000, respectively, angle of attack $\alpha=30.0°$, freestream temperature $T_\infty=270K$, adiabatic wall. Two simulations were performed, one with perfect gas and one with equilibrium real gas assumption. Shocks again are captured. In Fig. 13 the isobars for the Re=130,000 case for perfect gas on the left-hand-side and real gas on the right-hand-side show a main difference in real and perfect gas simulations. Due to the lower temperatures behind the shock in real gas flow the distance between the shock and the body is smaller than in perfect gas flow. The same can be seen in Fig. 14, where the isotherms are given for the same case. Note that real gas stagnation region temperature is about half of that for perfect gas. To study the influence of the Reynolds number a second perfect gas simulation with a Reynolds number of Re=13,000,000 was performed. On the leeward side of the vehicle the effect of the viscosity is smaller as was expected, see the temperature distribution in Fig. 15. From the skin friction pattern and from streamlines one can find that the flow separates on the wing, in front of the canopy and on top of the vehicle.

CONCLUSIONS

The NSFLEX method was proven to give reliable results for sub-, trans-, super- and hypersonic flow cases for perfect and real equilibrium viscous flows. The method will be applied especially in the hypersonic flow regime for example to the HERMES or the SÄNGER configuration. Present work is under progress to incorporate radiation boundary conditions, catalytic surfaces and slip. In future the code will be expanded to non-equilibrium real gas.

Acknowledgement: Thanks to the IABG - Ottobrunn VP200 supercomputer group for their support.

REFERENCES

[1] Schmatz, M.A.: Three-dimensional viscous flow simulations using an implicit relaxation scheme. In: Kordulla, W.(ed.): Numerical simulation of compressible viscous flow aerodynamics. Notes on Numerical Fluid Mechanics. Vieweg, 1988.

[2] Schmatz, M.A., Brenneis, A., Eberle, A.: Verification of an implicit relaxation method for steady and unsteady viscous and inviscid flow problems. AGARD Symp. Validation of CFD, Lisbon, 1988.

[3] Schmatz, M.A., Monnoyer, F., Wanie, K.M., Hirschel, E.H.: Zonal solutions of three-dimensional viscous flow problems. In: Zierep, J., Oertel, H. (eds.): Symposium Transsonicum III. Springer, 1988.

[4] Schmatz, M.A., Monnoyer, F., Wanie, K.M.: Numerical simulation of transonic wing flows using a zonal Euler/boundary-layer/Navier-Stokes approach. ICAS-paper 88-4.6.3, 1988.

[5] Eberle, A., Schmatz, M.A., Schaefer, O.: High-order solutions of the Euler equations by characteristic flux averaging. ICAS-paper 86-1.3.1, 1986.

[6] Eberle, A.: 3D Euler calculations using characteristic flux extrapolation. AIAA-paper 85-0119, 1985.

[7] Anderson, W.K., Thomas, J.L.. van Leer, B.: A comparison of finite volume flux vector splittings for the Euler equations. AIAA-paper 85-0122, 1985.

[8] Steger, J.L., Warming, R.F.: Flux vector splitting of the inviscid gasdynamic equations with application to finite difference methods. J. Comp. Phys., Vol. 40, 1981, pp. 263-293.

[9] Haenel, D., Schwane, R.: An implicit flux-vector splitting scheme for the computation of viscous hypersonic flow. AIAA-paper 89-0274, 1989.

[10] Srinivasan, S., Tannehill, J.C., Weilmünster, K.J.: Simplified curve fits for the thermodynamic properties of equilibrium air. ISU-ERI-Ames-88401 ERI project 1626 CFD 15, 1986.

[11] Eberle, A.: Characteristic flux averaging approach to the solution of Euler's equations. VKI lecture series, Computational fluid dynamics, 1987-04, 1987.

[12 Srinivasan, S., Tannehill, J.C., Weilmünster, K.J.: Simplified curve fits for the transport properties of equilibrium air. ISU-ERI-Ames-88405, ERI project 1626 CFD 21, 1987.

[13] Mundt, C., Keraus, R., Fischer, J.: paper in preparation, 1989.

[14] Somers, D.M.: Design and experimental results for a flapped natural-laminar-flow airfoil for general aviation applications. NASA-TP-1865, 1981.

[15] Heiss, S.: private communications, 1989.

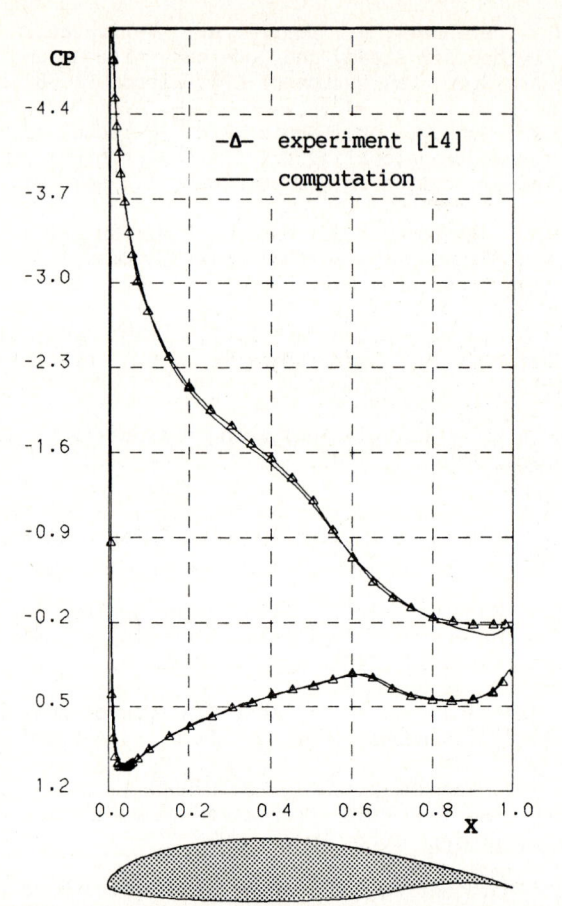

Fig. 4 Pressure distribution
(M$_\infty$=0.1, Re=6,000,000, α=13.21°)

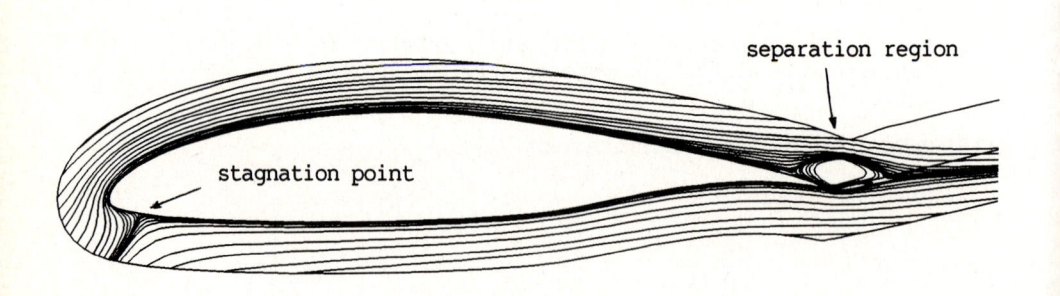

Fig. 5 Streamlines
(M$_\infty$=0.1, Re=6,000,000, α=13.21°)

shock position

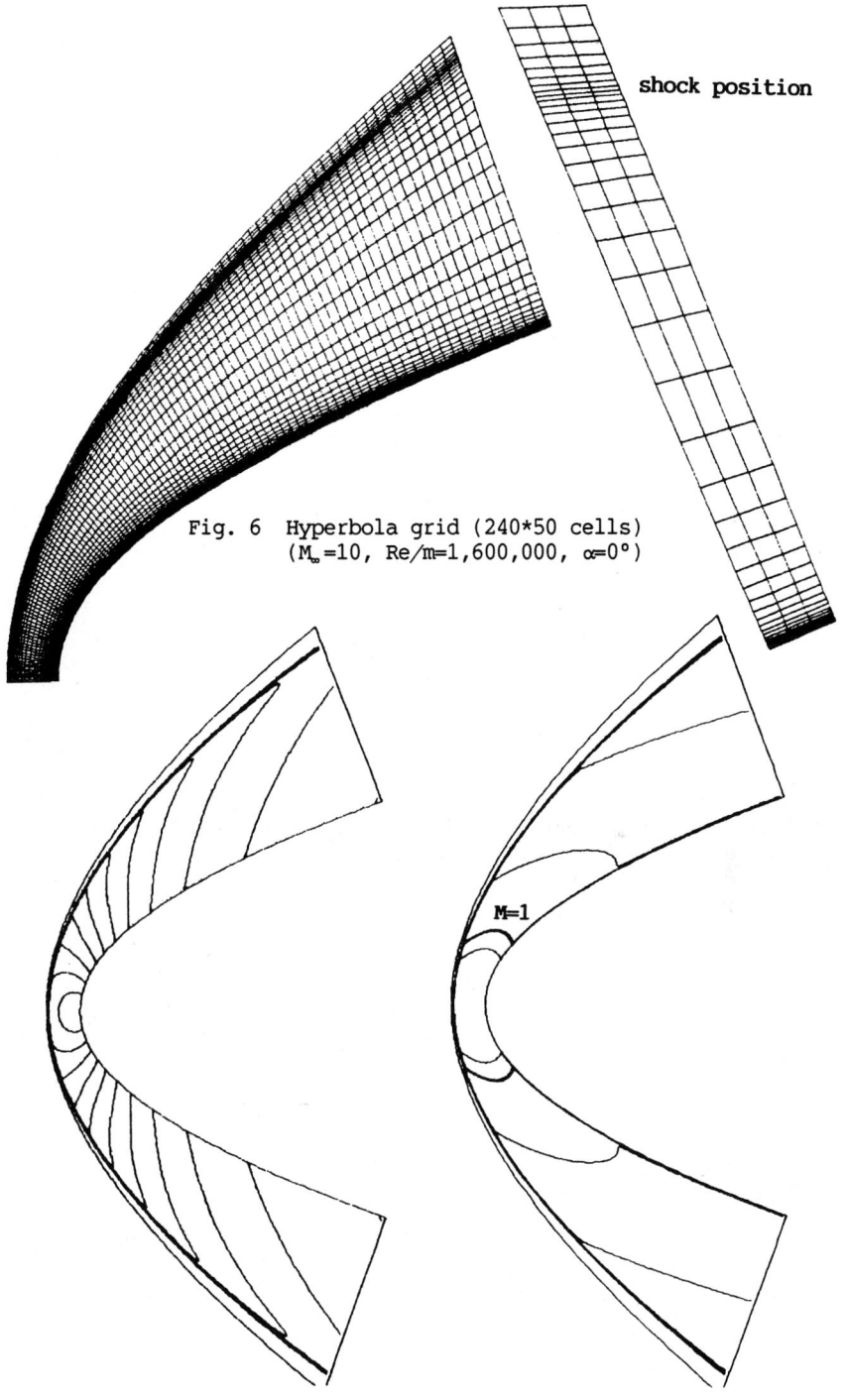

Fig. 6 Hyperbola grid (240*50 cells)
(M_∞=10, Re/m=1,600,000, α=0°)

Fig. 7 Isobars
(M_∞=10, Re/m=1,600,000, α=0°)

Fig. 8 Isotherms
(M_∞=10, Re/m=1,600,000, α=0°)

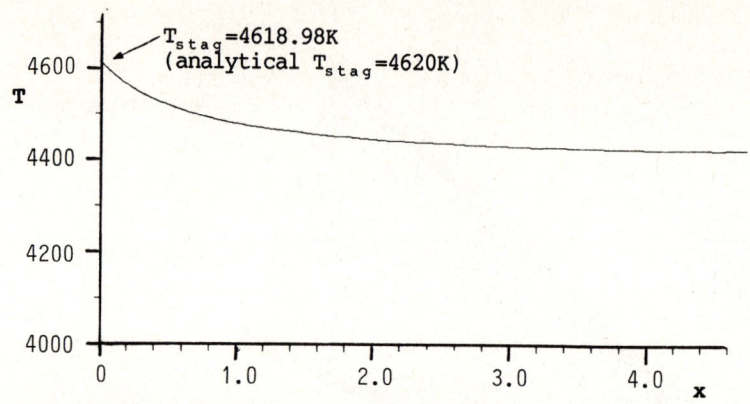

Fig. 9 Temperature distribution along body axis
($M_\infty=10$, Re/m=1,600,000, $\alpha=0°$)

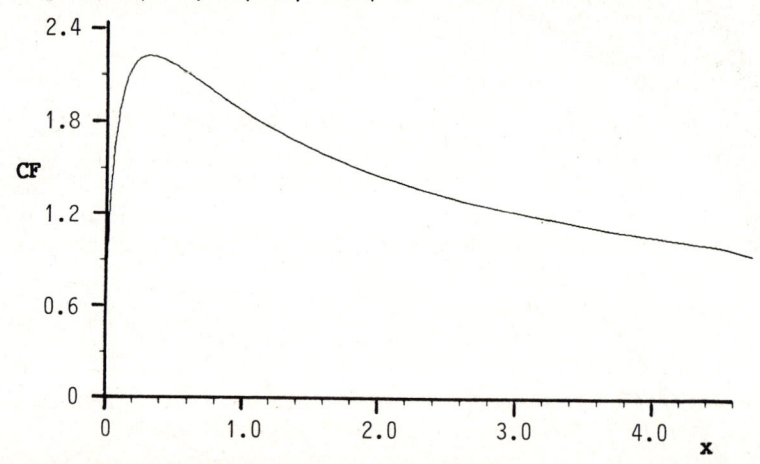

$T_{stag}=4618.98K$
(analytical $T_{stag}=4620K$)

Fig. 10 Skin friction distribution along body axis
($M_\infty=10$, Re/m=1,600,000, $\alpha=0°$)

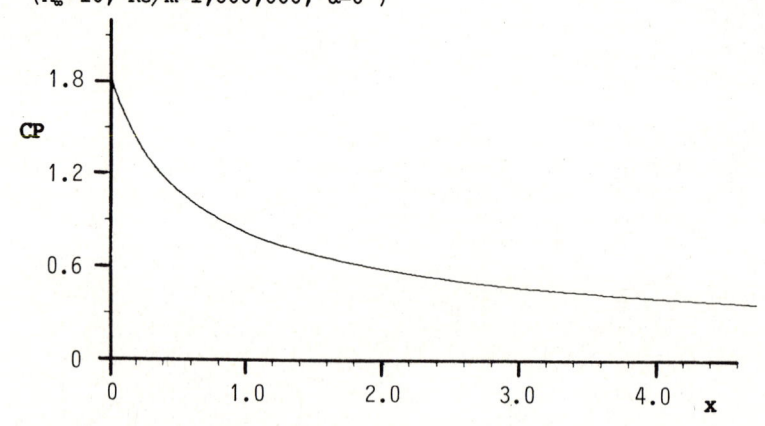

Fig. 11 C_p distribution along body axis
($M_\infty=10$, Re/m=1,600,000, $\alpha=0°$)

Fig. 12 Grid for HERMES vehicle

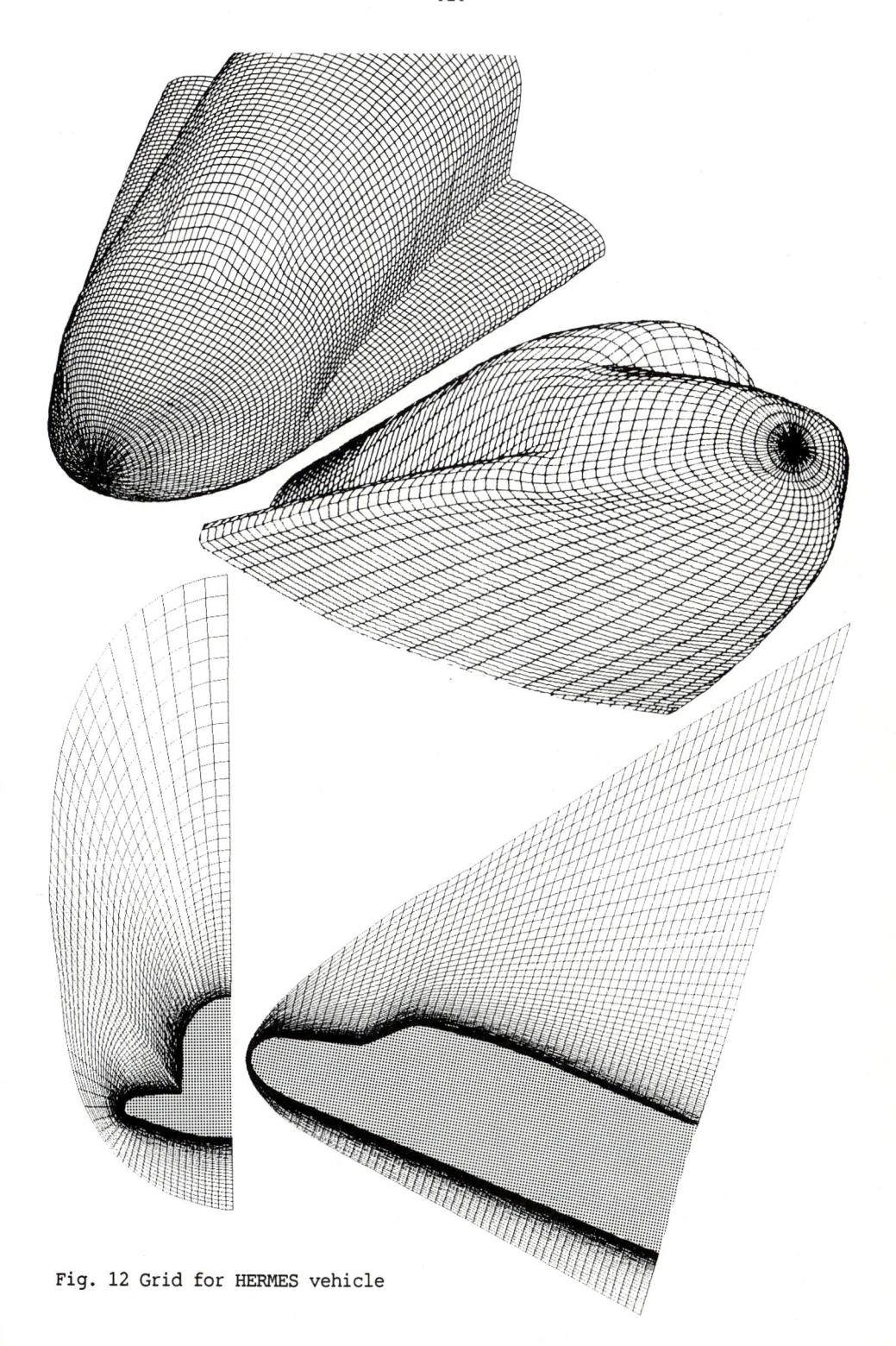

Fig. 13 Isobars
   (M_∞=10, Re=130,000, α=30°)

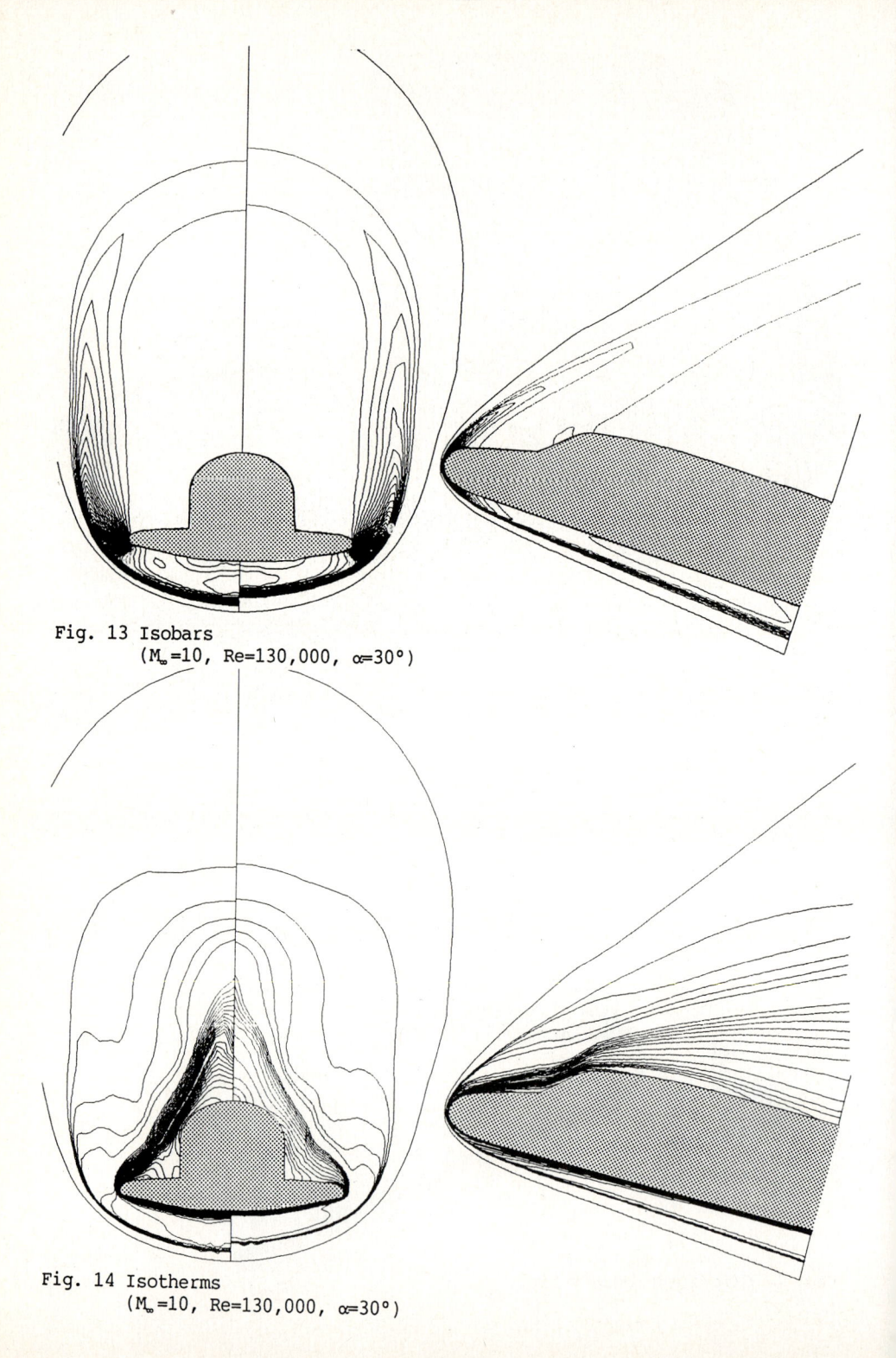

Fig. 14 Isotherms
   (M_∞=10, Re=130,000, α=30°)

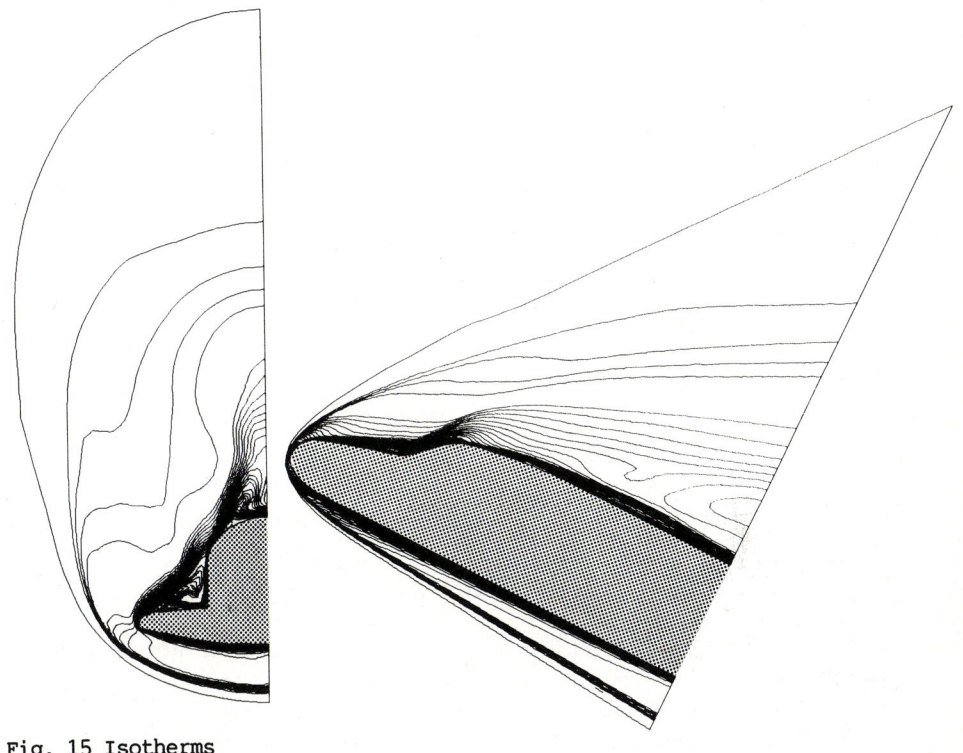

Fig. 15 Isotherms
($M_\infty=10$, Re=13,000,000, $\alpha=30°$)

A Multigrid Algorithm for the Incompressible Navier–Stokes Equations*

A. Schüller
Gesellschaft für Mathematik und Datenverarbeitung
F1/T
Postfach 1240
D–5205 St. Augustin 1

Abstract

In this paper a reformulation of the incompressible Navier–Stokes problem is derived, in which the continuity equation is replaced by a Poisson-like equation for the pressure. The corresponding discrete nonlinear system of equations has some advantageous features for the numerical solution. Results computed by a suitable multigrid algorithm are shown and discussed.

1 Introduction

The numerical calculation of realistic, complex fluid flow phenomena often requires more computational power than currently available. During the last two decades the efficiency of monoprocessor computers has increased by orders of magnitude. Physical constraints make similar improvements unlikely in the near future. On the other hand, parallel machines consisting of hundreds or thousands of processors have been designed and developed. Their computational power increases proportionally to the number of processors. Within a few years there will be a number of them available which have a peak performance of more than 100 Gflops. This will make possible the numerical solution of increasingly complicated problems.

Large systems of equations resulting from the discretization of differential equations are generally well suited for parallel computation. The reason for this is that these equations have some inherent locality: the equation for one unknown contains only values of unknowns which are located at neighbouring points. Thus, the main idea for the parallel solution of such problems is the partitioning of the domain on which the differential equations have to be solved. Each processor is responsible for the solution in a small part of the domain. Of course, for the solution of the whole problem a certain amount of communication between the processors is necessary.

In [8] these principles of parallelization are discussed in more detail. As an example, the parallelization of a multigrid algorithm for the solution of the Poisson equation is

---
*This article covers only one part of the lecture "Multigrid Methods for General Navier–Stokes Solution" given at the fifth GAMM-Seminar in Kiel by J. Linden, G. Lonsdale, H. Ritzdorf, A. Schüller, B. Steckel and K. Stüben. For more information see [10, 11]. All this work was done within the German supercomputer project SUPRENUM.

described there. The parallelization of a multigrid solver for the Navier–Stokes equations can be done in a similar way because the communication constructs and needs, which are essential for the parallel application, are very alike. For more information on a parallel multigrid solver of the Navier–Stokes equations on general 2D-domains see [11].

In this paper we discuss first some problems arising in the development of multigrid solvers for the incompressible Navier–Stokes equations, keeping in mind easy parallelizability and vectorizability of the algorithm. We assume that the reader is familiar with the basics of multigrid methods [3, 5, 16].

In section 3 we reformulate the incompressible Navier–Stokes equations such that some of the problems can be overcome. The new system automatically contains an elliptic term Δp in the equation, which replaces the original continuity equation. To guarantee full equivalence between the two systems of equations (provided all functions are sufficiently smooth) the continuity equation has to hold at the boundary of the domain.

In section 4 we discuss the correct multigrid treatment for the new boundary value problem. Some numerical results are given in section 5.

2 Some Problems in Solving the Incompressible Navier–Stokes Equations

In the development of multigrid solvers for the incompressible Navier–Stokes equations (2.1–2.3) one has to face a lot of difficulties even on conventional monoprocessor computers.

$$-\Delta u + Re(uu_x + vu_y) + p_x = f^u \qquad (2.1)$$
$$-\Delta v + Re(uv_x + vv_y) + p_y = f^v \qquad (2.2)$$
$$u_x + v_y = 0 . \qquad (2.3)$$

2.1 Grid and Discretization

First of all, for flows around complex geometries a suitable grid has to be chosen. Cartesian grids make the proper treatment of boundary conditions rather difficult. Moreover, many modern grid generation programs delivering boundary–fitted grids can generally create grids which are structured in blocks. Each of these blocks is a logically rectangular grid. This makes the data structures of the solution program, vectorization and parallelization much easier.

Using a finite volume discretization on such grids there are different ways to arrange the variables with respect to the grid. According to [10] a non–staggered arrangement (nodal point scheme) is better suited for general non–orthogonal grids than a staggered scheme. The accuracy of the computed results of staggered schemes depends much more sensitively on the non–orthogonality of the grid. Therefore we use the non–staggered scheme in the rest of this paper.

2.2 Stabilization by an Artificial Pressure Term

For simplicity we assume that we have a square cartesian grid. Moreover, let Dirichlet boundary conditions for u and v be given. Then the standard discretization of (2.1–

2.3) with central, second order finite differences produces the well known instability in p: oscillating functions for the pressure p are solutions of the homogeneous difference equations. Stabilization can be achieved by adding an artificial elliptic term

$$-\omega h^2 \Delta p \qquad (2.4)$$

to the continuity equation (2.3). Such artificial pressure terms are a common tool, implicitly also used by the SIMPLE algorithm described in [12, 13] and by some finite element approximations [1, 4] (mini–elements using so called "bubble functions"). Of course, one has to be careful that ω is not chosen too large. This would cause unphysical solutions. On the other hand, if ω is chosen too small, the oscillations in p are not damped enough and can still be seen in the solution. Moreover, the factor h^2 in the artificial pressure term causes a different scaling of different quantities with respect to h. In a multigrid algorithm with standard grid transfer components like injection or full weighting and linear interpolation this has the effect that the convergence rate of a 2–grid cycle is limited by 0.75 if ω is (much) too large and independent of h.

2.3 Boundary Conditions for p

Often boundary conditions for the velocities u and v are given. Then one needs an additional boundary condition for the pressure p with the non–staggered arrangement of variables. Constant extrapolation of values of p to the boundary can be interpreted as a first order discretization of

$$p_n = 0 \qquad (2.5)$$

which is a reasonable boundary condition for many flow problems. Linear extrapolation can be interpreted as a discretization of the boundary condition

$$p_{nn} = 0. \qquad (2.6)$$

As even the discrete Poisson equation $\Delta_h p_h = 0$ with this type of boundary condition has several linearly independent solutions, this condition has to be used with care, particularly if the artificial pressure term plays a major role. This may happen on grids with a rather large value of h.

2.4 Smoothing

In order to use multigrid to solve the standard discrete form (i. e. second order) of the equations (2.1–2.3), stabilized by (2.4), a smoothing scheme is required. In [11] a modified box–relaxation is proposed. Its smoothing factor predicted by local mode analysis is 0.625 for the Stokes problem ($Re = 0$) and $\omega = 1/16$.

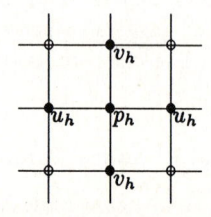

Figure 2.1: Variables updated by the smoothing algorithm for the nodal point scheme

In this scheme, which works similar to a smoothing scheme for the staggered arrangement of variables [17], each of the discrete unknowns u_h and v_h has to be updated twice within one smoothing step. Moreover, one smoothing step requires the solution of a 5×5 sparse system in every interior grid point (Figure 2.1). Therefore, this smoothing algorithm is somewhat expensive.

3   A Reformulation of the Incompressible Navier–Stokes Equations

Consider the incompressible Navier–Stokes equations (2.1–2.3). As explained in the previous section, many difficulties in the numerical solution of this system are caused by the special form of the continuity equation. There are several well known approaches to overcome this problem.

In their famous article describing the MAC method Harlow and Welch [6] obtain an equation containing Δp by differentiating the time–dependent momentum equations and adding them. Continuity on every new time level is guaranteed by solving this equation with an appropriate right hand side.[1] This method has been generalized to other time dependent problems by Hirt and Harlow [7].

There are a number of authors using similar approaches [14, discussion in chapter III–G]. Mostly, continuity is maintained by special techniques corresponding to the iteration in time. Obviously, this is not the most efficient approach to solve steady state problems.

Therefore, we derive a new system of equations which is equivalent to (2.1–2.3) and which can be solved more efficiently.

3.1   The Stokes Equations

First we consider the linear case of the Stokes equations.

Theorem: *Assume that Ω is a bounded domain in $I\!R^2$ and that $u(x,y)$ and $v(x,y)$ belong to $C_3(\Omega) \cap C_1(\overline{\Omega})$, $p(x,y) \in C_2(\Omega)$, $f^u(x,y) \in C_1(\Omega)$ and $f^v(x,y) \in C_1(\Omega)$. Then the two systems*

$$
\begin{align}
-\Delta u + p_x &= f^u \quad (\Omega) \tag{3.1}\\
-\Delta v + p_y &= f^v \quad (\Omega) \tag{3.2}\\
u_x + v_y &= 0 \quad (\overline{\Omega}) \tag{3.3}
\end{align}
$$

and

$$
\begin{align}
-\Delta u + p_x &= f^u \quad (\Omega) \tag{3.4}\\
-\Delta v + p_y &= f^v \quad (\Omega) \tag{3.5}\\
\Delta p &= f^u{}_x + f^v{}_y \quad (\Omega) \tag{3.6}\\
u_x + v_y &= 0 \quad (\partial\Omega) \tag{3.7}
\end{align}
$$

are fully equivalent.

---

[1] The continuity equation has to be valid on the boundary, too.

Proof: The sum of the derivatives of equations (3.1) with respect to x and (3.2) with respect to y is (3.6) as
$$-\Delta(u_x + v_y) = 0 \qquad (3.8)$$
because of (3.3).

In the other direction, equation (3.3) can be regained from equations (3.4–3.7) because the difference of the sum of the derivatives of equations (3.4) with respect to x and (3.5) with respect to y and equation (3.6) is equation (3.8) and this partial differential equation with the homogeneous Dirichlet boundary condition (3.7) interpreted as a boundary value problem for $u_x + v_y$ has the unique solution (3.3). $\square$

3.2 The Navier–Stokes Equations

This result can be generalized to the incompressible Navier–Stokes equations.

Theorem: *Assume that Ω is a bounded domain in $I\!R^2$ and that $u(x,y)$ and $v(x,y)$ belong to $C_3(\Omega) \cap C_1(\overline{\Omega})$, $p(x,y) \in C_2(\Omega)$, $f^u(x,y) \in C_1(\Omega)$ and $f^v(x,y) \in C_1(\Omega)$. Then the two systems*

$$\begin{aligned}
-\Delta u + Re(uu_x + vu_y) + p_x &= f^u & (\Omega) & \qquad (3.9) \\
-\Delta v + Re(uv_x + vv_y) + p_y &= f^v & (\Omega) & \qquad (3.10) \\
u_x + v_y &= 0 & (\overline{\Omega}) & \qquad (3.11)
\end{aligned}$$

and

$$\begin{aligned}
-\Delta u + Re(vu_y - uv_y) + p_x &= f^u & (\Omega) & \qquad (3.12) \\
-\Delta v + Re(uv_x - vu_x) + p_y &= f^v & (\Omega) & \qquad (3.13) \\
\Delta p + 2Re(v_x u_y - u_x v_y) &= f^u_{\ x} + f^v_{\ y} & (\Omega) & \qquad (3.14) \\
u_x + v_y &= 0 & (\partial\Omega) & \qquad (3.15)
\end{aligned}$$

are fully equivalent.

Proof: Because of (3.11) the momentum equations (3.9–3.10) can be written as (3.12–3.13). Differentiating equation (3.12) with respect to x and equation (3.13) with respect to y, adding these equations and using again the continuity equation (3.11) we get (3.14).

In the other direction, we regain continuity from equations (3.12–3.15) by the same arguments as in the Stokes case. $\square$

The problems in the numerical solution of the original Navier–Stokes system do not carry over to this boundary value problem as the "new continuity equation" (3.14) now naturally contains the elliptic term Δp.

3.3 Boundary Conditions

Two boundary conditions are necessary for the solution of the original Navier–Stokes equations (2.1–2.3). The new system (3.12–3.14) requires three, one of them being the continuity equation (3.15). There arises a new difficulty in the numerical solution of the

the new system. This has the consequence that its discrete analogue does not hold exactly but only up to some error caused by the discretization. Table 1 shows the dependence of $(u_x)_h + (v_y)_h$, measured in the discrete L_2 and maximum norms for $Re = 20$, on the mesh size of the discretization h. The results indicate that $(u_x)_h + (v_y)_h$ is computed with second order accuracy because asymptotically the errors decrease by a factor of about 4.

Table 1: Deviations from discrete continuity dependent on h

h	maximum norm	L_2 norm
1/4	.48	.15
1/8	.36	.81(−1)
1/16	.12	.22(−1)
1/32	.32(−1)	.57(−2)
1/64	.82(−2)	.15(−2)
1/128	.20(−2)	.37(−3)

6 Concluding Remarks

A system of differential equations has been derived, which is fully equivalent to the incompressible Navier–Stokes equations. For this system a multigrid algorithm is proposed, the smoothing scheme of which is a collective point relaxation. This algorithm proves to be very efficient for low and moderate values of Re. Moreover, it can be vectorized and parallelized easily.

The generalization to time dependent flows is straightforward.

An implementation of this algorithm for non–cartesian grids and for more general boundary conditions remains to be done.

7 Acknowledgements

I thank my colleagues R. Hempel, J. Linden, G. Lonsdale, H. Ritzdorf, B. Steckel and K. Stüben for their support and for many fruitful discussions.

References

[1] Abdalass, E. M.: *Resolution performante du probleme de Stokes par minielements, maillages auto–adaptifs et methodes multigrilles – applications.*, Ph. D. Thesis, Ecole Centrale de Lyon, 1987.

[2] Börgers, C.: *Mehrgitterverfahren für eine Mehrstellendiskretisierung der Poissongleichung und für eine zweidimensionale singulär gestörte Aufgabe.* Diplomarbeit, Institut für Angewandte Mathematik, Universität Bonn, 1981.

[3] Brandt, A.: *Guide to multigrid development*, in "Lecture Notes in Math.", 960, Springer, Berlin, 1982.

These are 6 equations for 6 unknowns (u_h, v_h and p_h at a boundary point and u_h, v_h and p_h at the corresponding auxiliary point).

If one wants to relax a point P on the boundary, it is not sufficient to do a collective update of these 6 quantities. The three equations in the grid point of $\overset{\circ}{\Omega}_h$, which is neighboured to P have to be included in this step of the relaxation in order to achieve good multigrid convergence. As the equations for u_h and v_h on the boundary are trivial, these unknowns can be eliminated and along the boundaries 7×7-systems have to be solved in every boundary point within a smoothing step.

Near corners, values for u_h and v_h at auxiliary points are calculated by extrapolation of $(u_0)_h$ and $(v_0)_h$. The three unkown values of p_h are determined by the equations (4.3–4.5) at the corner points.

5   Numerical Results

For all numerical results reported in this section the following parameters and functions describing the flow problem are given:

$$\Omega = (0,1)^2; \; v_0(x,y) = 0; \; f^u(x,y) = 0; \; f^v(x,y) = 0; \; u_0(x,y) = \begin{cases} 0 & \text{if } y < 1 \\ 16x^2(1-x)^2 & \text{if } y = 1 \end{cases}$$

The multigrid algorithm uses F-cycles, the full approximation scheme, linear interpolation of coarse grid corrections, two smoothing steps before and one after the coarse grid correction. The mesh size is 1/128 on the finest grid and 1/4 on the coarsest one.

The smoothing rate of the collective relaxation described in section 4.2 is 0.5 for the Stokes problem if the grid points are ordered lexicographically. This gives the estimation of 0.5^3 for the convergence rate per cycle. This is in very good agreement with values of about 0.11, which are measured for the multigrid solver.

Relaxation with lexicographical ordering of grid points is not suited for vectorization and parallelization. Therefore we implemented the same collective smoothing procedure with red–black ordering of grid points. This algorithm can be vectorized and parallelized very easily. The measured multigrid convergence rates using this smoother are about 0.10. This is slightly better than for the lexicographic case.

For the Navier–Stokes problem the discrete system and the convergence rates depend on the Reynolds number Re. For Re less than about 50 the multigrid algorithm using the smoothing algorithm with red–black ordering of grid points has convergence rates between 0.10 and 0.12.

If Re becomes too large, the multigrid algorithm starts to diverge as the relaxation on the coarsest grid diverges. This is not surprising because the ellipticity of the system degrades for $Re \to \infty$. Moreover, the ellipticity of the discrete system becomes worse the more h increases. The introduction of artificial viscosity (which is necessary on coarse grids only if Re is not too large) again gives a convergent multigrid algorithm. But it is well known that artificial viscosity worsens the convergence rates of multigrid [2]. This carries over to our multigrid algorithm.

Explicitly, the continuity equation $u_x + v_y = 0$ appears only as a boundary condition in where

$$\Delta_h = \begin{bmatrix} & \frac{1}{h_y^2} & \\ \frac{1}{h_x^2} & -\frac{2}{h_x^2} - \frac{2}{h_y^2} & \frac{1}{h_x^2} \\ & \frac{1}{h_y^2} & \end{bmatrix} \qquad (4.6)$$

and, e. g.,

$$(u_x)_h = \frac{1}{2h_x}\begin{bmatrix} -1 & 0 & 1 \end{bmatrix} u_h. \qquad (4.7)$$

4.2 Smoothing

The construction of a good smoothing algorithm is much simpler for the new Navier–Stokes problem than for the old one. The main part of each difference equation is characterized by the Δ_h-operator. The three equations are coupled by lower order terms. So, a smoothing method seems to be suitable which updates the three variables u_h, v_h and p_h by a collective point relaxation solving 3×3–systems in every grid point.

On a square grid such an algorithm has smoothing rates of 0.5 for the Stokes problem. This is better than the result for the box–relaxation which is a smoothing scheme suited for the original Navier–Stokes equations. Moreover, the amount of work per smoothing step is considerably lower for the collective relaxation.

4.3 Numerical Treatment of Boundary Conditions

The situation that a boundary value problem for a system of equations has several boundary conditions for some unknowns and none for others is not new. One well known example of this is the first boundary value problem for the biharmonic equation on a domain Ω

$$\Delta\Delta u = f \quad (\Omega) \qquad (4.8)$$
$$u = g \quad (\partial\Omega) \qquad (4.9)$$
$$u_n = \overline{g} \quad (\partial\Omega). \qquad (4.10)$$

As it is not easy to find good and efficient smoothing algorithms for direct discretizations of the biharmonic operator, the biharmonic equation is split into a system

$$\Delta v = f \quad (\Omega) \qquad (4.11)$$
$$\Delta u - v = 0 \quad (\Omega). \qquad (4.12)$$

If this system is discretized by standard second order finite differences on a cartesian grid, collective point relaxation is a suitable smoothing scheme. Obviously, this boundary value problem has two boundary conditions for u but none for v. Proper multigrid treatment for this problem is described in [9]. Some modifications improving the convergence properties of the V–cycle and applications to other problems such as shell equations can be found in [15].

It is possible to transfer these results to our problem. Along the boundary $\partial\Omega_h$ of Ω_h auxiliary points are required. The discrete equations on the boundary $\partial\Omega_h$ are (4.3–4.5) and

$$u_h = u_{0h} \qquad (4.13)$$
$$v_h = v_{0h} \qquad (4.14)$$
$$(u_x)_h + (v_y)_h = 0. \qquad (4.15)$$

problem if two boundary conditions for the velocity components u and v such as

$$g_1(u, v, u_n, v_n) = 0 \tag{3.16}$$
$$g_2(u, v, u_n, v_n) = 0 \tag{3.17}$$

are given. Then we have three conditions for the velocity components but no condition for the pressure p. What to do in such a case will be demonstrated by an example in the following section.

4 Numerical Treatment of the New System

In this section we discuss the solution of the system (3.12–3.14) on the domain $\Omega = [0, A] \times [0, B]$ with the boundary conditions (3.15) and

$$u = u_0 \tag{4.1}$$
$$v = v_0. \tag{4.2}$$

p is determined only up to a constant in this boundary value problem just as it is in the original Navier–Stokes problem. The treatment of such a singularity within multigrid is described in [5, chapters 11.4 and 12].

4.1 Discretization

The domain Ω is covered by a rectangular $N \times M$ point grid, Ω_h, with $h_x = A/(N-1)$, $h_y = B/(M-1)$, $h = (h_x, h_y)$. Let $\overset{\circ}{\Omega}_h$ denote the set of interior grid points of Ω_h.

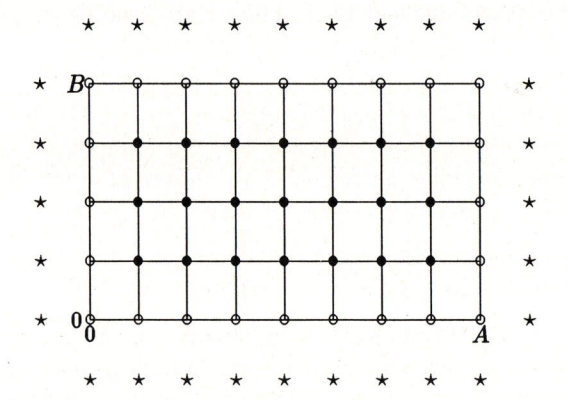

Figure 4.1: Grid points of $\partial \Omega_h$ are denoted by o, those of $\overset{\circ}{\Omega}_h$ by • and auxiliary points by ⋆.

Then in $\overset{\circ}{\Omega}_h$ the discrete equations can be written as

$$-\Delta_h u_h + Re(v_h(u_y)_h - u_h(v_y)_h) + (p_x)_h = f^u{}_h \tag{4.3}$$
$$-\Delta_h v_h + Re(u_h(v_x)_h - v_h(u_x)_h) + (p_y)_h = f^v{}_h \tag{4.4}$$
$$\Delta_h p_h + 2Re((v_x)_h(u_y)_h - (u_x)_h(v_y)_h) = (f^u{}_x + f^v{}_y)_h \tag{4.5}$$

[4] Brezzi, F.; Pitkäranta, J.: *On the stabilization of finite-element approximations of the Stokes equations*, in "Efficient Solution of Elliptic Systems" (Hackbusch, W., ed.), Notes on Numerical Fluid Mechanics 10, Vieweg, 1984.

[5] Hackbusch, W.: *Multigrid methods and applications*, Springer, Berlin, 1985.

[6] Harlow, F. H.; Welch, J. E.: *Numerical calculation of time–dependent viscous incompressible flow of fluid with free surface*, Physics of Fluids, Vol. 8, No. 12, 1965, pp. 2182–2189.

[7] Hirt, C. W.; Harlow, F. H.: *A general corrective procedure for the numerical solution of initial–value problems*, J. of Comp. Phys., Vol. 2, 1967, pp. 114–119.

[8] Hempel, R.; Schüller, A.: *Experiments with parallel multigrid algorithms using the SUPRENUM communications library*, GMD–Studien Nr. 141, Gesellschaft für Mathematik und Datenverarbeitung, St. Augustin, 1988.

[9] Linden, J.: *A multigrid method for solving the biharmonic equation on rectangular domains*, Arbeitspapiere der GMD 143, Gesellschaft für Mathematik und Datenverarbeitung, St. Augustin, 1985.

[10] Linden, J.; Lonsdale, G.; Steckel, B.; Stüben, K.: *Multigrid for the steady–state incompressible Navier–Stokes equations; a survey*, Arbeitspapiere der GMD 322, Gesellschaft für Mathematik und Datenverarbeitung, St. Augustin, 1988.

[11] Linden, J.; Steckel, B.; Stüben, K.: *Parallel multigrid solution of the Navier–Stokes equations on general 2D–domains*, Arbeitspapiere der GMD 294, Gesellschaft für Mathematik und Datenverarbeitung, St. Augustin, 1988.

[12] Peric, M.: *A finite volume method for the prediction of three–dimensional fluid flow in complex ducts*, Ph. D. Thesis, Mech. Eng. Dept., Imperial College, London, 1985.

[13] Rhie, C. M.; Chow, W. L.: *A numerical study of the turbulent flow past an isolated airfoil and trailing edge separation*, AIAA–82–0998.

[14] Roache, P. J.: *Computational Fluid Dynamics*, Hermosa Publishers, Albuquerque, New Mexico, 1976.

[15] Schüller, A.: *Mehrgitterverfahren für Schalenprobleme*, Dissertation, University of Bonn, 1987, published as GMD–Bericht Nr. 171, Gesellschaft für Mathematik und Datenverarbeitung, St. Augustin, 1988.

[16] Stüben, K.; Trottenberg, U.: *Multigrid methods: fundamental algorithms, model problem analysis and applications*, in "Lecture Notes in Math.", 960, Springer, Berlin, 1982.

[17] Vanka, S. P.: *Block–implicit multigrid solution of Navier–Stokes equations in primitive variables*, J. Comp. Phys. 65, 1986, pp. 138–156.

Analysis and Application of A Line Solver for the Recirculating Flows Using Multigrid Methods

T. M. Shah, D. F. Mayers, J. S. Rollett
Oxford University Computing Laboratory,
8-11 Keble Road, Oxford ENGLAND.

Abstract

This paper describes an efficient line solver for the coupled equations. The calculation procedure, called SCGS/LS(Symmetrical Coupled Gauss-Seidel/Line Solver), is based on the SCGS scheme. The technique is applied to the driven cavity problem for the incompressible Navier-Stokes equations. A Fourier analysis of this technique is carried out. The scheme is simple and easy to programme. The rates of convergence and the computational times are reported for the test case.

1 Introduction

In [11], a calculation procedure based upon a coupled simultaneous relaxation of the momentum and continuity equations has been successfully used in association with the multigrid method to calculate the flow fields in a driven cavity. The relaxation procedure used, is called symmetrical coupled Gauss-Seidel(SCGS) and it solves the equations node by node in a coupled manner. In a series of papers [10, 12, 2], its efficacy has been demonstrated on a number of standard test problems. A Fourier analysis of the SCGS scheme has been carried out in [5]. The analysis shows the good smoothing capabilities of the SCGS scheme.
In [7], a comparison of the smoothing capabilities of SCGS and SIMPLE has been presented. Their comparison demonstrated that SCGS resulted in a saving in execution time over SIMPLE. However, at the higher Reynolds numbers (5000,10000) and the finest mesh level (66 × 66) both methods converged at similar speed. This led us to think of an efficient line solver for the coupled equations.
The purpose of this paper is to present a line-solver (LS) based on the SCGS scheme. A Fourier analysis has also been carried out for the proposed method (SCGS/LS). The method is applied to the driven cavity problem. Calculations are made with finite-difference grids consisting of up to (130 × 130) grid nodes and the results are compared with earlier numerical studies.

2 Differential and Difference Equations

The equations expressing conservation of mass and momentum in two dimensions for an ideal, incompressible, Newtonian fluid are given by:

$$\frac{\partial \rho u^2}{\partial x} + \frac{\partial \rho uv}{\partial y} = -\frac{\partial p}{\partial x} + \frac{\partial}{\partial x}(2\mu \frac{\partial u}{\partial x}) + \frac{\partial}{\partial y}[\mu(\frac{\partial u}{\partial y} + \frac{\partial v}{\partial x})] \tag{1}$$

$$\frac{\partial \rho v^2}{\partial y} + \frac{\partial \rho uv}{\partial x} = -\frac{\partial p}{\partial y} + \frac{\partial}{\partial y}(2\mu \frac{\partial u}{\partial y}) + \frac{\partial}{\partial x}[\mu(\frac{\partial u}{\partial y} + \frac{\partial v}{\partial x})] \tag{2}$$

$$\frac{\partial \rho u}{\partial x} + \frac{\partial \rho v}{\partial y} = 0 \tag{3}$$

$(x,y) \in \Omega$ where x, y denote the co-ordinate directions and u, v the components of the velocity in these directions; p is the pressure and ρ and μ denote the density and viscosity of the fluid (assumed to be constant).

A Mac-type staggered grid (see [4]) system was used to locate the flow variables. The velocities are stored on the cell faces and the pressure is stored at the cell centres. There are several methods of devising finite-difference approximations but to ensure that the scheme is conservative, a finite volume approach is adopted. Due to the staggering of the mesh three different types of control volume are required for the two momentum equations and the continuity equation in the interior region, with straightforward modifications near the boundaries. A detailed description is given in [8].

The discrete equations are derived by first integrating the differential equations over each control volume surrounding the location of the variable. They are then expressed for a unit volume for convenient use of the multigrid technique. The fluxes over each control volume surface are constant.

The resulting finite-difference set of algebraic equations(for grid dimensions dx and dy) at each finite difference node (i,j) are as follows:

$$A_c^u u_{i+1/2,j} = A_n^u u_{i+1/2,j+1} + A_s^u u_{i+1/2,j-1} + A_e^u u_{i+3/2,j} \\ + A_w^u u_{i-1/2,j} + (p_{i,j} - p_{i+1,j})/\rho \delta x \qquad (4)$$

$$A_c^v v_{i,j+1/2} = A_n^v v_{i,j+3/2} + A_s^v v_{i,j-1/2} + A_e^v v_{i+1,j+1/2} \\ + A_w^v v_{i-1,j+1/2} + (p_{i,j} - p_{i,j+1})/\rho \delta y \qquad (5)$$

$$(u_{i+1/2,j} - u_{i-1/2,j})/\delta x + (v_{i,j+1/2} - v_{i,j-1/2})/\delta y = 0 . \qquad (6)$$

The superscripts relate to the variables and the subscript indices denote the locations of the variables on the finite difference grid. The fractional indices refer to the staggered locations of the velocities. The coefficients A_c^u, A_c^v, etc., are same as described in [5].

3   Solution Algorithm

3.1   Derivation of Equations for SCGS/LS

To derive equations along a paraxial line (x, or y axis) for the SCGS scheme, rewrite (4-6) for a finite difference node (i,j). The pressure $p_{i,j}$ is located at the centre and the velocities u, v are staggered halfway between the nodes in each direction (x, y). The derived equations are

$$(A_c^u)_{i-1/2,j} u_{i-1/2,j} = F_{i-1/2,j}^u \qquad (7)$$
$$(A_c^u)_{i+1/2,j} u_{i+1/2,j} = F_{i+1/2,j}^u \qquad (8)$$
$$(A_c^v)_{i,j-1/2} v_{i,j-1/2} = F_{i,j-1/2}^v \qquad (9)$$
$$(A_c^v)_{i,j+1/2} v_{i,j+1/2} = F_{i,j+1/2}^v \qquad (10)$$

and

$$(u_{i+1/2,j} - u_{i-1/2,j})/\delta x + (v_{i,j+1/2} - v_{i,j-1/2})/\delta y = 0 \qquad (11)$$

where, for example:

$$F^u_{i+1/2,j} = A^u_n u_{i+1/2,j+1} + A^u_s u_{i+1/2,j-1} + A^u_e u_{i+3/2,j}$$
$$+ A^u_w u_{i-1/2,j} + (p_{i,j} - p_{i+1,j})/\rho \delta x. \tag{12}$$

In terms of corrections and residuals along a paraxial line(say y-axis), (7-11) may be written as follows:

$$(A^u_c)_{i-1/2,j} u'_{i-1/2,j} + (p'_{i,j} - p'_{i-1,j})/\rho h = R^u_{i-1/2,j} \tag{13}$$
$$(A^u_c)_{i+1/2,j} u'_{i+1/2,j} - (p'_{i,j} - p'_{i+1,j})/\rho h = R^u_{i+1/2,j} \tag{14}$$
$$(A^v_c)_{i,j-1/2} v'_{i,j-1/2} + p'_{i,j}/\rho h = R^v_{i,j-1/2} \tag{15}$$
$$(A^v_c)_{i,j+1/2} v'_{i,j+1/2} - p'_{i,j}/\rho h = R^v_{i,j+1/2} \tag{16}$$
$$(u'_{i+1/2,j} - u'_{i-1/2,j})/h + (v'_{i,j+1/2} - v'_{i,j-1/2})/h = R^c_{i,j} \tag{17}$$

where primed variables are corrections; $h = \delta x = \delta y$ a uniform mesh size and $R^u_{i+1/2,j} = F^u_{i+1/2,j} - (A^u_c)_{i+1/2,j} u_{i+1/2,j}$, etc. $R^c_{i,j}$ is the LHS of (11).

Now solve (13-16) for u' and v' and substitute in (17) to get the required system of equations

$$u'_{i-1/2,j} = [R^u_{i-1/2,j} - (p'_{i,j} - p'_{i-1,j})/\rho h]/(A^u_c)_{i-1/2,j} \tag{18}$$
$$u'_{i+1/2,j} = [R^u_{i+1/2,j} + (p'_{i,j} - p'_{i+1,j})/\rho h]/(A^u_c)_{i+1/2,j} \tag{19}$$
$$v'_{i,j-1/2} = [R^v_{i,j-1/2} - p'_{i,j}/\rho h]/(A^v_c)_{i,j-1/2} \tag{20}$$
$$v'_{i,j+1/2} = [R^v_{i,j+1/2} + p'_{i,j}/\rho h]/(A^v_c)_{i,j+1/2} \tag{21}$$

and

$$(A^p_c)_{i+1,j} p'_{i+1,j} + (A^p_c)_{i,j} p'_{i,j} + (A^p_c)_{i-1,j} p'_{i-1,j} = R^p_{i,j}. \tag{22}$$

Coefficients A^p_c and residuals $R^p_{i,j}$ are given as follows:

$$(A^p_c)_{i+1,j} = -\frac{1}{\rho h^2 (A^u_c)_{i+1/2,j}}$$
$$(A^p_c)_{i-1,j} = -\frac{1}{\rho h^2 (A^u_c)_{i-1/2,j}}$$
$$(A^p_c)_{i,j} = [\frac{1}{(A^u_c)_{i+1/2,j}} + \frac{1}{(A^u_c)_{i-1/2,j}} + \frac{1}{(A^v_c)_{i,j+1/2}} + \frac{1}{(A^v_c)_{i,j-1/2}}]/\rho h^2$$

$$R^p_{i,j} = R^c_{i,j} - 1/h[R^u_{i+1/2,j}/(A^u_c)_{i+1/2,j} - R^u_{i-1/2,j}/(A^u_c)_{i-1/2,j}]$$
$$- 1/h[R^v_{i,j+1/2}/(A^v_c)_{i,j+1/2} - R^v_{i,j-1/2}/(A^v_c)_{i,j-1/2}].$$

Equation (22) with j fixed and $i = 1, 2, 3, ..., N$ are a triple diagonal set of equations for the $p'_{i,j}, p'_{i+1,j}, ... p'_{N,j}$ pressure variables. A tri-diagonal solver can be applied to (22) for each pressure correction; Substitution into (19-21) yields the corrections for u, v. Add these corrections to the latest calculated values of u, v, p along the same line and then move to the next j line, and so on until the domain Ω is completed. The alternative sweep direction requires the same amount of work by rewriting (13-16) for a line of constant x, the rest of the algorithm is the same.

3.2 Multigrid Methods

Briefly, the concept of using multigrid techniques arose because any unconverged solution to a set of discrete elliptic equations contains errors of a wide range of frequencies. Conventional methods like Gauss-Seidel, ADI, etc., are efficient in smoothing errors of wavelengths comparable to the mesh size, but their convergence for low frequency errors is quite slow. However, frequencies small on one grid are relatively large on a coarser grid. Therefore, cycling between the fine grid and a series of coarse grids can smooth errors of all frequencies in an optimal way. There are several ways of multigrid cycling. We used FAS, suitable for non-linear problems. A detailed review of multigrid methods and variant strategies can be found in [9] and [1].

3.3 Grid coarsening

There are several choices of grid coarsening like doubling the given mesh size, semi-coarsening, red-black, etc. We adopted a different grid coarsening called "continuity control volume lumping" (see [8]), which ensures that the compatibility continuity condition between the fine and the coarse grid is satisfied. In short, each coarse grid continuity control volume is composed of four fine grid continuity control volumes. This follows the discrete compatibility condition on all grids, which is crucial to obtaining optimal multigrid convergence rates.

3.4 Restriction and Prolongation

Restriction refers to the process of evaluating coarse grid residuals and corrections from fine grid values. Prolongation is the opposite process of obtaining fine grid values and corrections from coarse grid corrections. In the present study, the restricted coarse grid velocities are defined to be the means of their two nearest neighbouring fine grid velocities. Coarse grid pressures are defined to be the means of the four neighbouring fine grid pressures. The prolongation relations are derived by a bilinear interpolation. For the prolongation, four fine grid values are calculated from four adjacent coarse grid nodes; pressures have different weightings because of their cell-centred locations. Details of restrictions and prolongations can be found in [8].

3.5 Smoothing Procedure

The block implicit multigrid algorithm (BLIMM) used the SCGS scheme as a smoothing procedure (see [11, 10, 12]), where finite difference equations (7-11) on any grid are solved simultaneously by a point Gauss-Seidel method. The grid is scanned in a pre-determined manner; and for each continuity control volume, the momentum equations corresponding to the velocities on all four sides of the control volume and the continuity equation are solved in a coupled manner.

In the present study, we developed a line solver for the coupled system of equations. We solved the derived (22) for pressure corrections along a constant line, then a direct substitution into (19-21) yields values of the velocities along the same line. An under-relaxation parameter (ω_u, ω_v) was implemented by dividing the coefficients A_c^u, A_c^v respectively. The grid is scanned line by line in one direction followed by a similar sweep in the other direction. The sweeps are repeated, in the multigrid strategy, until the desired convergence is obtained on the finest grid. The convergence criterion is based on the summed averaged residuals in the three equations.

4 Fourier Analysis of SCGS/LS

4.1 Local Mode Analysis

To give an idea of the behavior of the relaxation schemes in iterative methods, local mode analysis is an efficient tool. Since the reduction of high frequency error components is essentially a local process, the analysis of this reduction need not take account of distant boundaries. Consider an arbitrary local section of the mesh with u, v velocities staggered along the x, y directions respectively and the pressure p located at nodes.

Assume that at the start of the smoothing process, errors in $u, v,$ and p are given (in Fourier modes) as follows

$$\begin{bmatrix} \delta u \\ \delta v \\ \delta p \end{bmatrix} = \begin{bmatrix} \delta u_0 \\ \delta v_0 \\ \delta p_0 \end{bmatrix} \exp(\iota \theta . x/h) \tag{23}$$

where $\theta . x/h = (\theta_1 x + \theta_2 y)/h$. Then, during the smoothing process, the singly corrected and the fully corrected errors are given by

$$\begin{bmatrix} \dot{\delta u} \\ \dot{\delta v} \end{bmatrix} = \begin{bmatrix} \delta u_1 \\ \delta v_1 \end{bmatrix} \exp(\iota \theta . x/h) \tag{24}$$

and

$$\begin{bmatrix} \ddot{\delta u} \\ \ddot{\delta v} \\ \ddot{\delta p} \end{bmatrix} = \begin{bmatrix} \delta u_2 \\ \delta v_2 \\ \delta p_2 \end{bmatrix} \exp(\iota \theta . x/h) \tag{25}$$

where $\dot{}$, $\ddot{}$ denote singly and fully corrected values. Notice that u and v are each corrected twice, but p only once. After some technical manipulation it can be shown that (see [5])

$$\begin{bmatrix} \delta u_1 \\ \delta v_1 \\ \delta u_2 \\ \delta v_2 \\ \delta p_2 \end{bmatrix} = \begin{bmatrix} M_1 \\ M_2 \end{bmatrix} \begin{bmatrix} \delta u_0 \\ \delta v_0 \\ \delta p_0 \end{bmatrix}, \tag{26}$$

where M_1 is a two by three complex matrix and M_2 is a three by three complex matrix. Thus the smoothing procedure can be written:

$$\begin{bmatrix} \delta u_2 \\ \delta v_2 \\ \delta p_2 \end{bmatrix} = M_2 \begin{bmatrix} \delta u_0 \\ \delta v_0 \\ \delta p_0 \end{bmatrix}. \tag{27}$$

The smoothing factor is then given by

$$\bar{\mu} = \sup_{\theta \in \mathcal{H}} [\rho(M_2)]$$

where $\mathcal{H} = [-\pi, \pi]^2 \setminus [-\pi/2, \pi/2]^2$ is the set of high frequencies. In the case of convection dominated flows, we use

$$\bar{\mu} = \sup_{|u_0, v_0| \leq 1, \ \theta \in \mathcal{H}} [\rho(M_2)]$$

where u_0, v_0 are frozen velocities used to linearize the problem and are constrained in order to maintain the relevant Reynolds number, $\rho(M_2)$ is the spectral radius of the matrix M_2.

4.2 Smoothing Rates

To obtain smoothing rates of the SCGS/LS, local mode analysis was carried out for the driven cavity problem. It is found that the underlying scheme has excellent smoothing capabilities especially at higher Reynolds numbers ($Re > 1000$). Both schemes (SCGS and SCGS/LS) have same smoothing capabilities at lower Reynolds numbers. Tables 1 and 2 give comparisons of the smoothing rates over a wide range of Reynolds numbers for the said schemes, together with the optimal relaxation parameters. The SCGS/LS scheme is, indeed, far more efficient than SIMPLE. This is because, for SCGS/LS, the practical smoothing rate remains well below the theoretical maximum at all Reynolds numbers. A comparison of SCGS and SIMPLE has been recently reported in [7]. A more detailed account of the theoretical smoothing capabilities of each of the schemes and plots of the reduction factors can be found in [5] and [6]. Here we give a few plots of the amplification factors of the SCGS/LS scheme.

Figs. 1-2 show contour plots of the amplification factors at various Reynolds numbers for a flow direction $(u_0, v_0) = (0, 1)$. The contour k gives a value $\mu(\theta) = k/10$. Fig.1 shows the amplification factor at Reynolds number 100. The smoothing factor in this case is $\bar{\mu} = .403$. At such a low Reynolds number the amplification factor is independent of flow direction. This indicates that the underlying line solver is a satisfactory smoother for diffusion dominated flows.

In Fig. 2 the information is depicted for convection dominated flow at Reynolds number 10000 for the same flow direction. It is illustrated in this figure that the amplification is largest along the line $\theta_2 = 0$, which is perpendicular to the flow direction $u_0 = 0$. The discrete ellipticity analysis (see [6]) also shows that the ellipticity is nearly lost along the line parallel to the θ plane which is perpendicular to the flow direction. This has a great effect on the amplification factor at high Reynolds numbers, which slows down the smoothing rates. However, in practice, the coupled methods efficiently enhance the smoothing rates.

5 Test of Solution Method

5.1 Test Problem

For the purpose of demonstrating the applicability and the robustness of the proposed scheme and to evaluate its performances relative to the earlier procedures [7], the driven cavity problem was solved. This test problem has been widely used for validating solution procedures for the Navier-Stokes equations. The underlying scheme is treated as a two stage process. After solving a triple diagonal system of equations for pressures, velocities can be calculated by a direct substitution. The velocities are updated twice and the pressures once.

In order to get a grid independent convergence rate, the grid is swept line by line in alternating directions. This is achieved cheaply at low Reynolds numbers but it is more expensive, at higher Reynolds numbers, as we have to employ symmetric alternating directions. Numerical experiments show that symmetric alternating directions do not accelerate the convergence much at low Reynolds numbers ($Re < 1000$). The same has been reported for the SCGS scheme. The convergence criterion is based on the summed average residuals in the three equations. i.e,

$$R_{con} = [\Sigma_{i,j}[(R^u_{i,j})^2 + (R^v_{i,j})^2 + (R^c_{i,j})^2]/(3 \times i_{max} \times j_{max})]^{1/2},$$

where R^u, R^v and R^c are residuals in the u, v momentum equations and in the continuity equations. i_{max}, j_{max} are the number of internal mesh points over which the summation is made. R_{con} was set to 10^{-4} and when the fine grid residual decreased below this value, the calculations were terminated. It is found to be most efficient to carry out just one iteration in each cycle on every grid except the coarsest one. We solved the equations, on the coarsest grid, exactly by performing more than one local iteration.

5.2 Results

Using the non-linear multigrid procedure and the proposed (SCGS/LS) algorithm as a smoother, the test problem has been solved over a range of Reynolds numbers from 1 to 10000. The results are compared with earlier work [7], on a typical mesh size. The results are extremely promising. Smoothing analysis (section 4.1) also shows that the underlying smoothing procedure is robust. To obtain a practical smoothing rate, the multigrid method is applied to solve the problem under consideration on the finest grid and to calculate the asymptotic convergence rate ρ_{mg}. The method used ν_1 pre- and ν_2 post relaxations. Hence the practical smoothing rate $\bar{\mu}_p$ is defined by

$$\bar{\mu}_p = (\rho_{mg})^{1/(\nu_1+\nu_2)}.$$

Tables 3 and 4 give values of the practical smoothing rate ($\bar{\mu}_p$) of both schemes for the test problem together with minimum and maximum theoretical smoothing rates. Fig. 3 shows graphically the behavior of the practical smoothing rates for SCGS/LS together with upper and lower bounds on the theoretical smoothing rates. It is clear that the convergence rate, at the higher Reynolds number is becoming constant in SCGS/LS, but this was not the case in SCGS (see [7]). Further at low Reynolds numbers, the smoothing factor is almost independent of the flow direction; hence the theoretical analysis cannot predict accurately enough the lower bound on the theoretical smoothing rates.

Tables 5 and 6 summarise the convergence characteristics of SCGS and SCGS/ LS and provide a comparison of their performance. These tables also show a strong indication that grid independent convergence is being approached at all Reynolds numbers. The numerical experiments also show that if the problem is solved accurately enough on the coarsest grid then the convergence rate does not deteriorate with mesh refinement. Tables 5 and 6 also demonstrate that each SCGS/LS iteration takes less CPU time than SCGS. This is because, when solving by line the coefficients A_c^u ,etc., need to be evaluated $3N$ times as opposed to $4N$ times in the case of a point solver, where N is the number of nodes in the x-direction.

Figs. 4-6 are plots of the streamfunction at Reynolds number 1, 100, 10000 on a 66×66 mesh, and are self-explanatory. The streamfunction is obtained by calculating the vorticity and then solving a Poisson equation.

Fig.7 is a plot of the streamfunction at Reynolds number 10000 on a (130×130) mesh, which indicates that the strength of the primary vortex increases with mesh refinement. This result is fairly close to that of Ghia et al [3].

6 Conclusion

The theoretical and practical smoothing capabilities of the proposed scheme (SCGS/LS) have been presented. The theoretical analysis shows that the underlying scheme has better h-independent convergence rates than SCGS at all Reynolds numbers, and in practice such convergence rates have been achieved. Numerical experiments demonstrated that such a solver is competitive in terms of execution time and simplicity of equations and coding, when compared to SIMPLE, SCGS and other schemes applied to complex incompressible flows. The scheme is robust in its insensitivity to the relaxation factors especially at the low Reynolds numbers. The behavior of the practical smoothing rates (Fig. 5) leads us to think that with a better choice of either multigrid strategy or some continuation technique, smoothing rates, at higher Reynolds numbers, may be improved further. So far, we have developed a continuation technique which needs to be extended further for the implementation of the underlying scheme.

Numerical experiments have shown that the underlying scheme converged monotonically even at $Re = 20000$. This shows the efficiency of the scheme; hence we suggest that it would be worthwhile to consider further test problems like rectangular cavities with backward and forward steps. These are currently under investigation.

Acknowledgement

One of the author (Shah) is grateful for the financial support of the government of Pakistan, and the encouragement of Professor K. W. Morton.

References

[1] A. Brandt. Guide to multigrid development. *Lecture Notes in Mathematics*, (960), 1981. Springer Verlag.

[2] P. H. Gaskell and N. G. Wright. Multigrids applied to an efficient fully coupled solution technique for Recirculating fluid flow problems. In *Simul. and Optimisation of large System*. IMA, IMA, september 1986.

[3] K. N. Ghia, U. Ghia and C. T. Shin. High-Re solution for incompressible flow using the Navier-Stokes equations and a multigrid method. *J. Comput. Phys.*, (48):387–411, 1982.

[4] F. H. Harlow and J. E. Welch. Numerical calculation of time-dependent viscous incompressible flow of fluid with free surface. *The Physics of Fluids*, 8(12), December 1965.

[5] T. M. Shah. Analysis of a multigrid method. Master's thesis, Oxford University, Oxford University computing Laboratory, 1987.

[6] G. J. Shaw and S. Sivaloganathan. On the smoothing properties of the SIMPLE pressure correction algorithm. *Inter. J. Num. Methods for Fluids*, 8:441–462, 1988.

[7] S. Sivaloganathan G. J. Shaw T. M. Shah and D. F. Mayers A comparison of multigrid methods for the incompressible Navier-Stokes equations. In K. W. Morton and M. J. Baines, editors, *Numerical Methods for Fluid Dynamics*. Oxford University Press, March 1988.

[8] S. Sivaloganathan and G. J. Shaw. A multigrid method for Recirculating flows. *Inter. J. Num. Methods for Fluids*, 8:417–440, 1988.

[9] K. Stuben and U. Trottenberg. Multigrid methods: Fundamental algorithms, model problem analysis and application. *Lecture Notes in Mathematics*, (960):1–176, 1981.

[10] S. P. Vanka. Block implicit multigrid calculation of two-dimensional Recirculating flows. *Comput. Meths. Appl. Mech. Engrg.*, (59):29–48, 1986.

[11] S. P. Vanka. Block implicit multigrid solution of the Navier-Stokes equations in primitive variables. *J. Comput. Phys.*, (65):138–158, 1986.

[12] S. P. Vanka. A calculation procedure for three-dimensional steady Recirculating flows using multigrid methods. *Comput. Meths. Appl. Mech. Engrg.*, (55):321–338, 1986.

Table 1: Relaxation Factors and Smoothing Factors (SCGS/LS)

Re	ω_u	ω_v	ω_p	$\bar{\mu}_{min}$	$\bar{\mu}_{max}$
1	1	1	1	0.352	0.352
100	1	1	1	0.332	0.431
400	0.7	0.7	1	0.409	0.586
1000	0.5	0.5	1	0.423	0.775
5000	0.4	0.4	1	0.435	0.911
10000	0.4	0.4	1	0.435	0.953

Table 2: Relaxation Factors and Smoothing Factors (SCGS)

Re	ω_u	ω_v	ω_p	$\bar{\mu}_{min}$	$\bar{\mu}_{max}$
1	0.7	0.7	1	0.323	0.323
100	0.7	0.7	1	0.329	0.415
400	0.6	0.6	1	0.339	0.520
1000	0.5	0.5	1	0.376	0.807
5000	0.5	0.5	1	0.392	0.990
10000	0.5	0.5	1	0.395	0.994

Table 3: Theoretical and Practical Smoothing Factors (SCGS/LS)

Re	$\bar{\mu}_{min}$	$\bar{\mu}_{max}$	$\bar{\mu}_p$
1	0.352	0.352	0.283
100	0.332	0.431	0.283
400	0.409	0.586	0.429
1000	0.423	0.775	0.600
5000	0.435	0.911	0.621
10000	0.435	0.953	0.631

Table 4: Theoretical and Practical Smoothing Factors (SCGS)

Re	$\bar{\mu}_{min}$	$\bar{\mu}_{max}$	$\bar{\mu}_p$
1	0.352	0.352	0.290
100	0.332	0.431	0.290
400	0.409	0.586	0.516
1000	0.423	0.775	0.600
5000	0.435	0.911	0.631
10000	0.435	0.953	0.641

Table 5: Number of Multigrid iterations required to reduce residual norm by 10^{-4}. (Figures in parentheses are CPU times on a DEC Microvax) (SCGS/LS)

Finest Grid Re	2 6x6	3 10x10	4 18x18	5 34x34	6 66x66
1 ($\omega = 1.0$)	4 (1s)	4 (3s)	4 (19s)	4 (1m 11s)	4 (4m 50s)
100 ($\omega = 1.0$)	5 (1s)	6 (5s)	5 (23s)	4 (1m 11s)	4 (4m 50s)
400 ($\omega = 0.7$)	6 (2s)	6 (5s)	8 (32s)	8 (2m 23s)	5 (6m 02s)
1000 ($\omega = 0.5$)	6 (2s)	9 (7s)	11 (42s)	11 (2m 46s)	10 (11m 54s)
5000 ($\omega = 0.4$)	7 (2s)	10 (8s)	9 (60s)	11 (6m 03s)	13 (24m 45s)
10000 ($\omega = 0.4$)	7 (2s)	11 (9s)	10 (83s)	13 (7m 08s)	15 (33m 06s)

Table 6: Number of Multigrid iterations required to reduce residual norm by 10^{-4}. (Figures in parentheses are CPU times on a DEC Microvax) (SCGS)

Finest Grid Re	2 6x6	3 10x10	4 18x18	5 34x34	6 66x66
1 ($\omega = 0.7$)	4 (1s)	5 (6s)	5 (24s)	5 (1m 20s)	4 (6m 07s)
100 ($\omega = 0.7$)	7 (2s)	5 (6s)	6 (26s)	6 (1m 36s)	4 (6m 07s)
400 ($\omega = 0.6$)	6 (1s)	7 (7s)	9 (36s)	9 (3m 00s)	7 (10m 42s)
1000 ($\omega = 0.5$)	8 (2s)	8 (9s)	9 (36s)	10 (3m 10s)	8 (17m 31s)
5000 ($\omega = 0.5$)	8 (2s)	9 (10s)	11 (44s)	18 (5m 20s)	18 (32m 18s)
10000 ($\omega = 0.5$)	10 (3s)	11 (12s)	14 (57s)	22 (6m 35s)	22 (45m 00s)

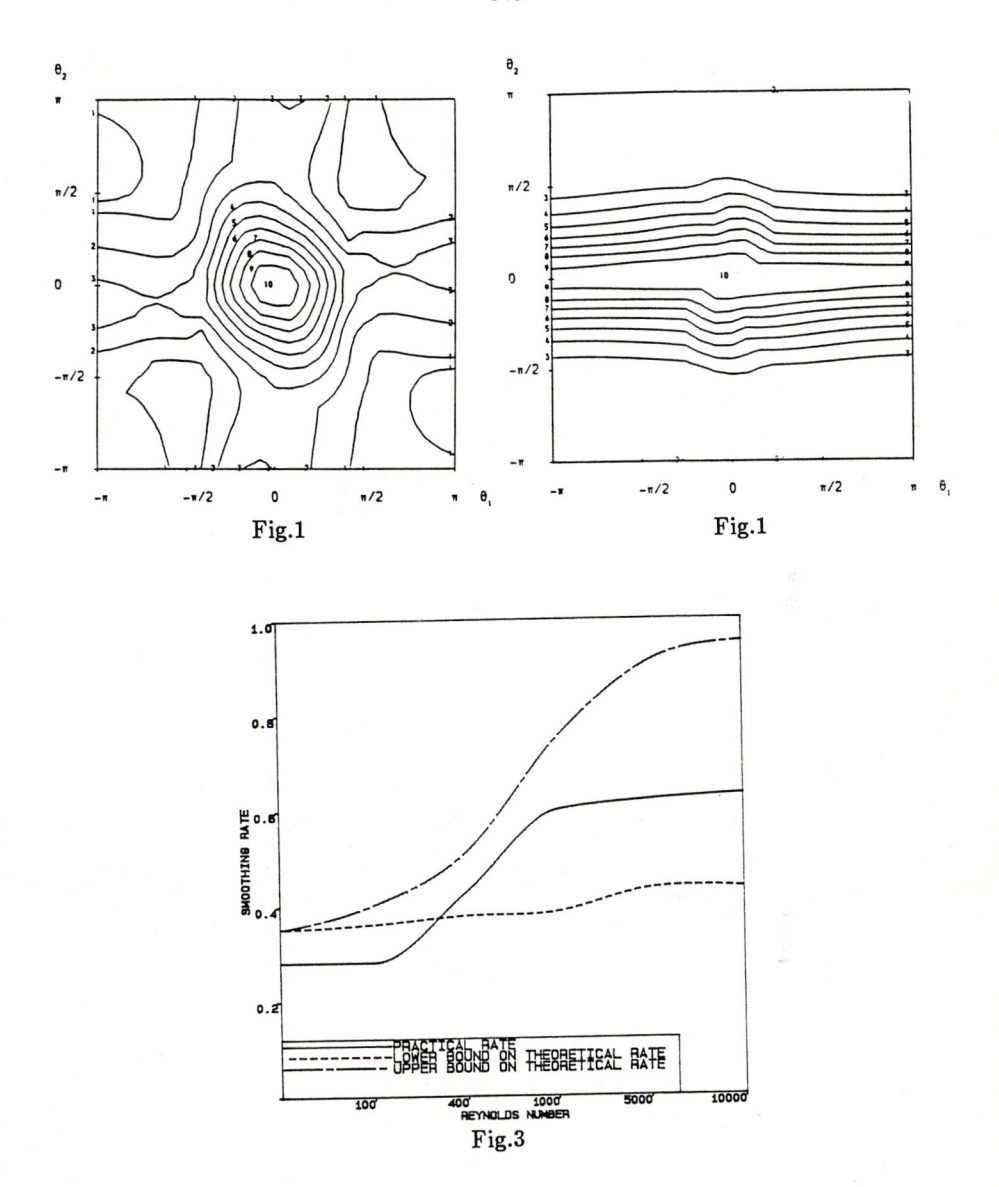

Fig.1 Amplification Factor at Reynolds Number 100 $(u_0, v_0) = (0, 1)$
Fig.2 Amplification Factor at Reynolds Number 10000 $(u_0, v_0) = (0, 1)$
Fig.3 Comparison of Theoretical and Practical Smoothing Rates: (SCGS/LS)

Fig.4

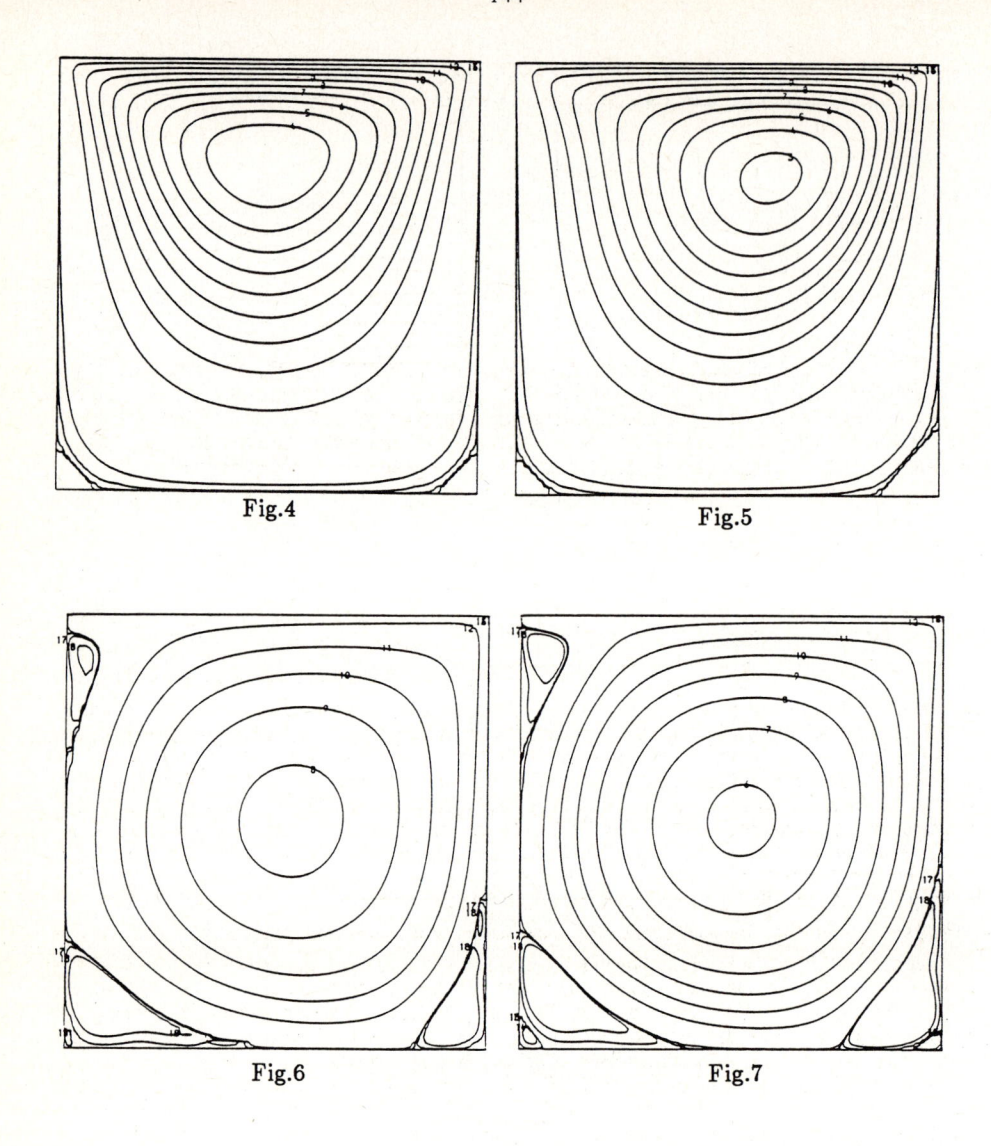

Fig.5

Fig.6

Fig.7

Fig.4 Streamfunction at Reynolds Number 1 Mesh size: (66x66) (SCGS/LS)
Fig.5 Streamfunction at Reynolds Number 100 Mesh size: (66x66) (SCGS/LS)
Fig.6 Streamfunction at Reynolds Number 10000 Mesh size: (66x66) (SCGS/LS)
Fig.7 Streamfunction at Reynolds Number 10000 Mesh size: (130x130) (SCGS/LS)

A Posteriori Error Estimators and Adaptive Mesh-Refinement for a Mixed Finite Element Discretization of the Navier-Stokes Equations

R. Verfürth
Institut für Angewandte Mathematik, Universität Zürich
Rämistr. 74, CH-8001 Zürich, Switzerland

SUMMARY

We present an a posteriori error estimator for mixed finite element approximations of the stationary, incompressible Navier-Stokes equations which is based on evaluating on each triangle the residual of the computed solution with respect to the strong form of the differential equation. Numerical examples for the Taylor-Hood element show the efficiency of adaptive mesh-refinement based on this a posteriori error estimator.

INTRODUCTION

The process of generating a sequence of triangulations using adaptive mesh-refinement based on a posteriori error estimators can be described as follows: starting with a coarse initial triangulation, one solves approximately the discrete problem corresponding to the actual triangulation, computes for each triangle an estimate of the error of the computed discrete solution and then, using this error estimate, decides which triangles have to be refined thus obtaining the next triangulation. Two problems apparently arise during this process: how to estimate the local error and how to decide which triangles have to be refined.

For linear elliptic equations, several a posteriori error estimators have been proposed in the literature. One possibility (cf. [2,3]) is to solve on a small patch of triangles Dirichlet problems of the same type as the original equation using appropriate higher order finite element spaces. Similarly, one can also solve on each triangle a Neumann problem (cf. [4]). Usually, this method leads to local problems of a smaller size. Another approach consists in using higher order finite differences in order to estimate locally the $W^{2,\infty}$-norm of the solution and in combining this information with a priori L^∞-error estimates (cf.[6]).

In [11], we generalized the approach of [4] to mixed finite element approximations of the Navier-Stokes equations. Here, we estimate the error of the discrete solution by evaluating locally its residual with respect to the strong form of the differential equation.

Different strategies for local mesh-refinement have been proposed in [2,3,4]. Here, we use a strategy which is based on the heuristic argument that among all triangulations with the same number of nodes that one is optimal for which all local errors are the same. Consequently, we divide all those triangles which have an estimated error above a certain threshold γ times the maximum of the estimated errors of all triangles.

DISCRETIZATION OF THE NAVIER-STOKES EQUATIONS

The steady-state motion of an incompressible fluid with velocity $\mathbf{u} = (u,v)$ and pressure p inside a bounded domain Ω in R^2 is described by the steady-state incompressible Navier-

Stokes equations

$$-\Delta \mathbf{u} + \mathrm{Re}(\mathbf{u}\cdot\nabla)\mathbf{u} + \nabla p = \mathbf{f} \text{ in } \Omega$$
$$\nabla\cdot\mathbf{u} = 0 \text{ in } \Omega \tag{1}$$
$$\mathbf{u} = 0 \text{ on } \Gamma := \partial\Omega$$

where Re $\geq$ 0 denotes the Reynolds' number. Note, that the restriction to homogeneous Dirichlet boundary conditions is made only in order to simplify the exposition. Non-homogeneous and mixed boundary conditions can be treated similarly.

Denote by $H^k(\Omega)$, $k\geq 0$, and $L^2(\Omega) := H^0(\Omega)$ the usual Sobolev and Lebesgue spaces (cf. [1]) equipped with the inner product

$$(u,v)_k := \sum_{|\alpha|\leq k} \int_\Omega D^\alpha u \, D^\alpha v$$

and the corresponding norm

$$\|u\|_k := (u,u)_k^{1/2} .$$

$H^1_0(\Omega)$ is the sub-space of all H^1-functions vanishing on Γ and $L^2_0(\Omega)$ is the sub-space of all L^2-functions having mean-value zero.

Let T_k, $k\geq 0$, be a family of triangulations of Ω which satisfies the following assumptions:
(T1) any two triangles may meet at most in a common vertex or a whole common edge,
(T2) the angles of all triangles in T_k are bounded away from 0 uniformly in k,
(T3) every triangle in T_k is either contained in T_{k-1} or is obtained by cutting a triangle of
T_{k-1} either into four new triangles by joining the midpoints of the edges or by connecting the midpoint of one edge to the vertex opposite to this edge (cf. Fig. 1).

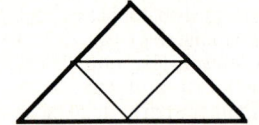

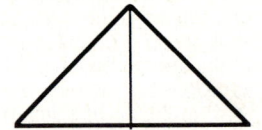

Figure 1: cutting a triangle into four and two new ones

Assumptions (T1) and (T2) are standard regularity hypotheses for triangulations, whereas (T3) ensures that the finite element spaces defined below are nested.

Denote by S_k the space of all continuous functions which are piecewise linear on the triangles in T_k and define

$$X_k := \{S_k \cap H^1_0(\Omega)\}^2 , \quad Y_k := S_{k-1} \cap L^2_0(\Omega).$$

Then we consider the Taylor-Hood approximation of problem (1) which is given by:

Find $\mathbf{u}_k \in X_k$, $p_k \in Y_k$ such that

$$(\nabla\mathbf{u}_k,\nabla\mathbf{v}_k)_0 + \mathrm{Re}((\mathbf{u}_k\cdot\nabla)\mathbf{u}_k,\mathbf{v}_k)_0 - (p_k,\nabla\cdot\mathbf{v}_k)_0 = (\mathbf{f},\mathbf{v}_k)_0 \quad \forall \mathbf{v}_k \in X_k \tag{2}$$
$$(\nabla\cdot\mathbf{u}_k,q_k)_0 = 0 \quad \forall q_k \in Y_k .$$

Note, that the spaces X_k and Y_k satisfy the so-called Babuska-Brezzi condition (cf. [5,9]), which guarantees the unique solvability of problem (2) for sufficiently small Reynolds' numbers (cf. [7]), provided each triangle in T_k is obtained by dividing a triangle of T_{k-1} into four new ones. If this assumption is not satisfied, oszillations in the pressure may theoretically occur. Our numerical results, however, show that in practice this is not the case even for strong local mesh-refinements.

AN ERROR ESTIMATOR BASED ON LOCAL RESIDUALS

Let $T \in T_k$ be a triangle and E be an edge of T. Denote by |T| and |E| the area of T and the length of E, respectively. Let $\mathbf{u}_k \in X_k$, $p_k \in Y_k$ be a solution of problem (2). Then the error estimator for triangle T is defined by

$$\eta_T^2 := |T| \, \|\mathbf{f}+\Delta\mathbf{u}_k - Re(\mathbf{u}_k \cdot \nabla)\mathbf{u}_k - \nabla p_k\|_{L^2(T)}^2 + |E| \sum_{E \in \partial T \cap \Omega} \|[\frac{\partial \mathbf{u}_k}{\partial n}]_J\|_{L^2(E)}^2 + \|\nabla \cdot \mathbf{u}_k\|_{L^2(T)}^2 \quad (3)$$

where $[\frac{\partial \mathbf{u}_k}{\partial n}]_J$ denotes the jump of $\frac{\partial \mathbf{u}_k}{\partial n}$ across the edge E. The proof of the following theorem is given in [11] for the mini-element approximation of the Navier-Stokes equations together with the modifications which are necessary for the Taylor-Hood element.

THEOREM: Assume that the assumptions (T1)-(T3) hold and that the Reynolds' number is sufficiently small. Then there exist two constants $0 < c_1 \leq c_2$, which only depend on Ω, the smallest angle in T_k, and the Reynolds' number, such that the following estimates hold for the unique solutions $\mathbf{u}$, p and $\mathbf{u}_k$, p_k of problem (1) in its standard weak form and of problem (2), respectively:

$$\|\mathbf{u}-\mathbf{u}_k\|_1 + \|p-p_k\|_0 \leq c_1 \{ \sum_{T \in T_k} \eta_T^2 \}^{1/2} \quad (4)$$

$$\eta_T \leq c_2 \{\|\mathbf{u}-\mathbf{u}_k\|_{H^1(T)} + \|p-p_k\|_{L^2(T)} + O(|T|)\} \quad \forall \, T \in T_k. \quad (5)$$

Note, that the proof of the theorem does not require any regularity assumptions on the solution of problem (1) which are stronger than those needed for the formulation of the weak problem. The assumption of small Reynolds' numbers is a technical one which is needed for a fixed point argument.

A MESH-REFINEMENT STRATEGY

As indicated in the introduction, our mesh-refinement strategy is based on the following rule:

(R0) a triangle $T \in T_k$ is cut into four new triangles if $\eta_T \geq \gamma \max \{\eta_\tau : \tau \in T_k\}$.

The parameter γ may be arbitrarily chosen in the interval (0,1). Values of γ close to 1 yield a sharp local refinement, whereas values close to 0 produce a nearly uniform refinement. In our calculations we always chose $\gamma = 0.5$.

In order to obtain triangulations satisfying assumption (T1), the mesh-refinement must obey to the following two additional rules:

(R1) a triangle having at least two neighbours cut into four new triangles must also be cut into four new ones,
(R2) a triangle having exactly one neighbour cut into four new triangles must be cut into two new ones by joining the midpoint of the edge, which has been divided, with the vertex opposite to this edge.

Here, two triangles are said to be neighbours if they share a common edge.

Finally, we have to add the following rule, which guarantees that the triangulations satisfy the minimal angle condition (T3):
(R3) if a predecessor of a triangle T has been cut into two new triangles, triangle T can only be cut into four new ones.

The resulting refinement process is illustrated in figure 2. There the numbers refer to the rules which underly the refinement of the different triangles.

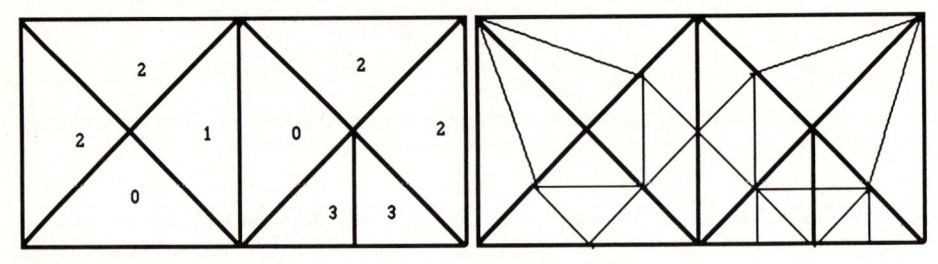

Figure 2: refinement of triangles according to the rules (R0) - (R3)

NUMERICAL RESULTS

The numerical tests of this section were performed on an Apple Mac Intosh II™ using an Absoft Mac Fortran™ compiler version 2.4. The discrete nonlinear problems are solved with a nonlinear least squares conjugate gradient algorithm as described in [8]. In each nonlinear iteration three discrete Stokes problems must be solved. This is done by a conjugate gradient version of the Uzawa algorithm as described in [10]. At each linear iteration step two Poisson-type equations must be solved. This is done approximately using a multi-grid algorithm kindly provided by G. Wittum. In order to accelerate the solution process, we only solve the discrete problems on a given level if the number of unknowns has increased at least by a factor of two with respect to the last level on which the discrete problems were solved. Otherwise, we simply interpolate the approximate solution from the preceeding level.

Our first example is the well-known driven cavity at Reynolds' number 50. Figure 3 shows the triangulation after 9 refinement steps together with zooms of the upper left and upper right corners of the square. It demonstrates that the mesh-refinement is concentrated in the upper corners of the square where the solution is singular and that the refinement process also takes account of the assymmetry of the solution.

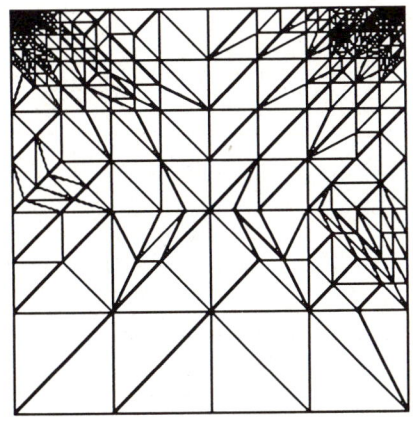

Figure 3a: driven cavity at Reynolds' number 50, level 9, 872 triangles

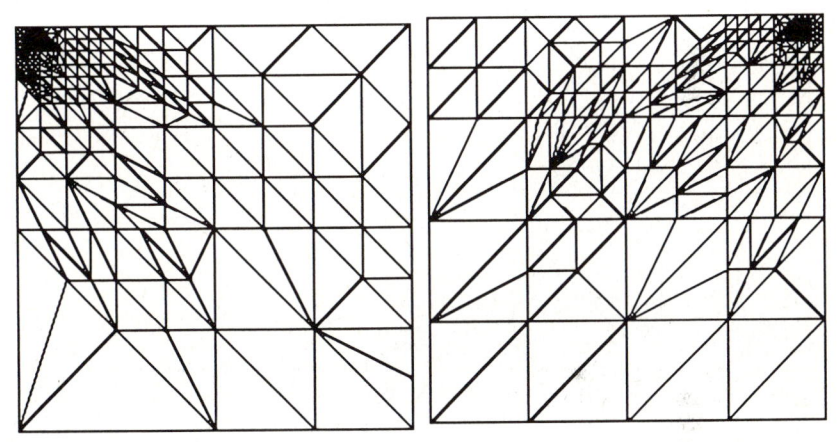

Figure 3b: zoom of (0.0,0.25)x(0.75,1.0)   Figure 3c: zoom of (0.75,1.0)x(0.75,1.0)

The next example is the flow across a backward facing step. The channel length/hight upstream and downstream of the step are 3/1 and 19/2 units, respectively. At the in- and outflow parabolic velocity profiles are prescribed. The maximum velocity at the inflow is 1. The corresponding Reynolds' number is 5. Figure 4 shows the triangulation after 7 refinement steps together with two zooms of the region near the step. The additional refinement at the outflow is due to the simple refinement strategy. It can be avoided by choosing a slightly larger value for γ.

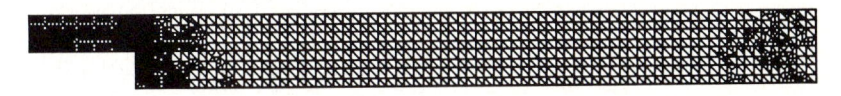

Figure 4a: backward facing step at Reynolds' number 5, level 7, 4430 triangles

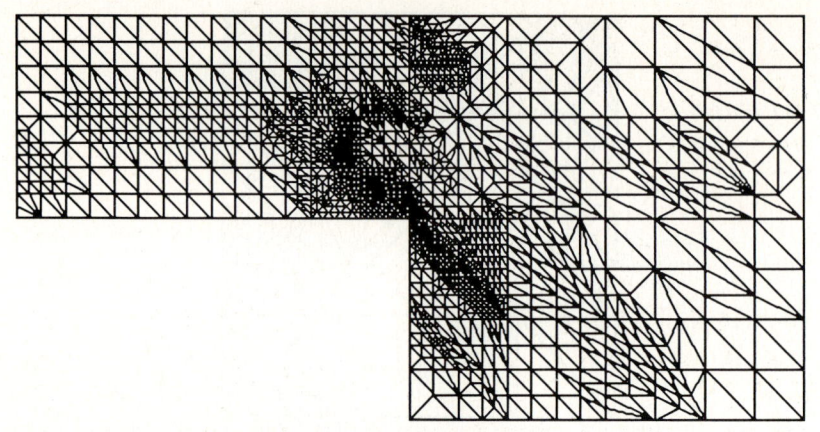

Figure 4b: zoom of the region (-2.0,2.0)x(-1.0,1.0)

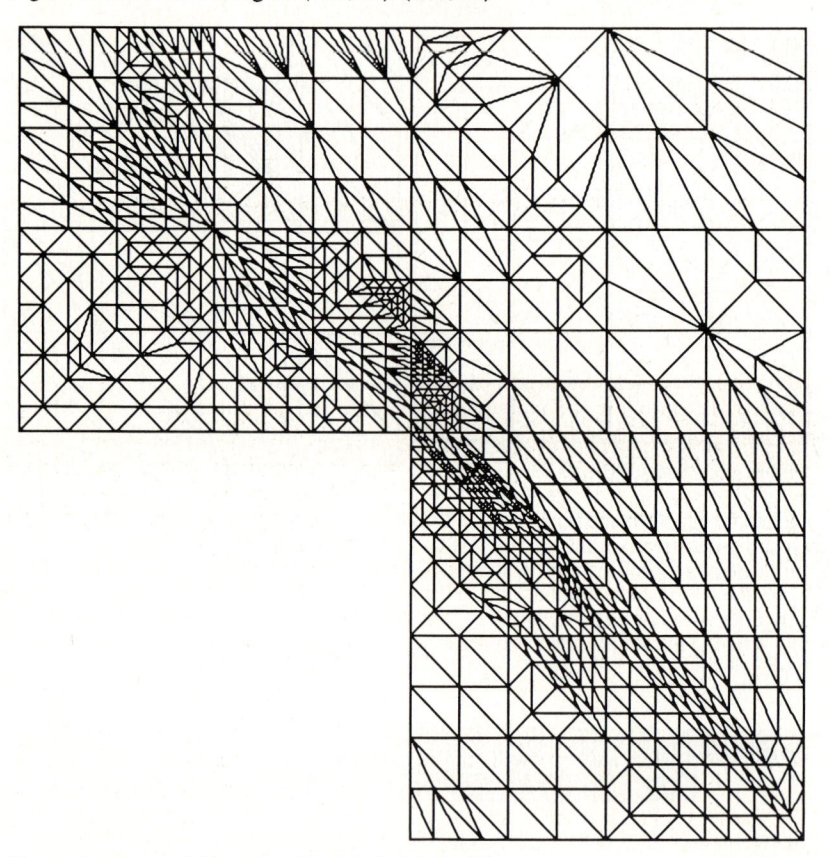

Figure 4c: zoom of the region (-0.5,0.5)x(-0.5,0.5)

As a final example we consider the Stokes flow, i.e. Re = 0, in a circular segment of radius 1 and angle 270°. The exact solution is of the form

$$\mathbf{u}(r,\varphi) = r^\alpha \psi(\varphi) \;,\; p(r,\varphi) = r^{\alpha-1}\chi(\varphi)$$

with $\alpha \approx 0.544484$ and suitable functions ψ and χ. Figure 5 shows the initial triangulation together with a partial view of the velocity field near the re-entrant corner. Figure 6 presents the triangulation after 6 refinement steps together with a zoom of the region near the re-entrant corner.

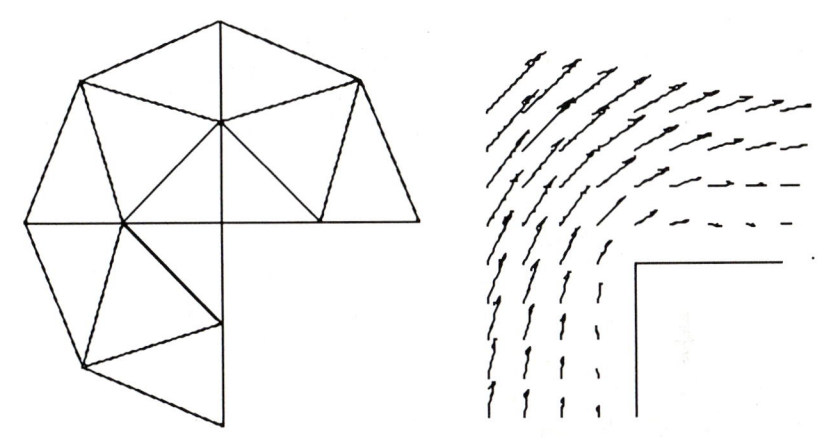

Figure 5: Stokes flow in a cicular segment; initial triangulation and velocity field near the corner

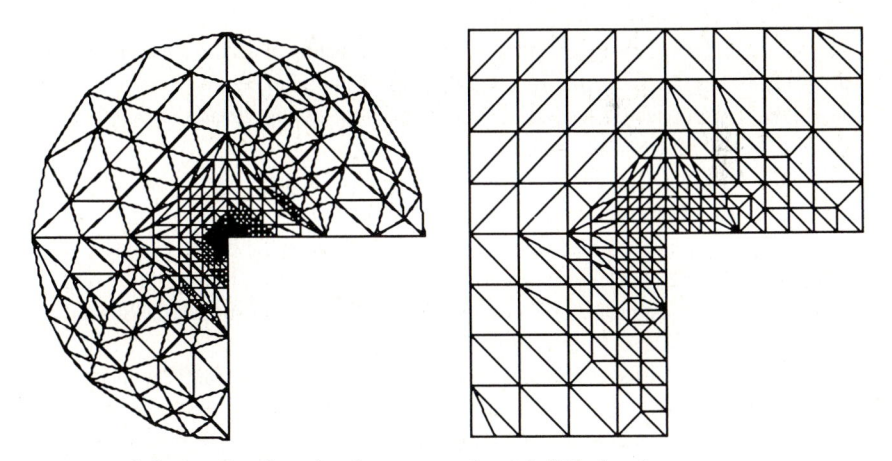

Figure 6: Stokes flow in a circular segment, level 6, 554 triangles

In table 1, we finally give the number of triangles (NT), the number of unknowns (NN), the relative error e_u of the velocity in the H^1-norm, the relative error e_p of the pressure in the L^2-norm, and the computing times in seconds for generating the triangulations (TT) and for solving the dicrete problems (TS) both for uniform refinement (4 levels) and local refinement (6 levels). It clearly underlines the efficiency of the error estimator and of the refinement process.

Table 1: uniform and local refinement for Stokes flow in a circular segment

	NN	NT	e_u	e_p	TT	TS
uniform refinement	3072	3341	3.8	7.4	9.5	333.8
local refinement	554	663	4.2	8.1	8.4	106.3

REFERENCES

[1] R. A. Adams : Sobolev spaces. Academic Press, New York 1975
[2] I. Babuska, W. C. Rheinboldt : A posteriori error estimates for the finite element method. Int. J. Numer. Methods in Engrg. 12, 1597-1615 (1978)
[3] I. Babuska, W. C. Rheinboldt : Error estimates for adaptive finite element computations. SIAM J. Numer. Anal. 15, 736-754 (1976)
[4] R. E. Bank, A. Weiser : Some a posteriori error estimators for elliptic partial differential equations. Math. Comput. 44, 283-301 (1985)
[5] F. Brezzi : On the existence, uniqueness, and approximation of saddle-point problems arising from Lagrangian multipliers. RAIRO Anal Numér. 8, 129-151 (1974)
[6] K. Eriksson, C. Johnson: An adaptive finite element method for linear elliptic problems. Math. Comput. 50, 361-383 (1988)
[7] V. Girault, P. A. Raviart : Finite element approximation of the Navier-Stokes equations. Computational Methods in Physics, Springer, Berlin 1986
[8] R. Glowinski : Numerical Methods for Nonlinear Variational Problems. Springer, Berlin, 1984
[9] R. Verfürth : Error estimates for a mixed finite element approximation of the Stokes equations. RAIRO Anal. Numér. 18, 175-182 (1984)
[10] R. Verfürth : A combined conjugate gradient - multi-grid algorithm for the numerical solution of the Stokes problem. IMA J. Numer. Anal. 4, 441-455 (1984)
[11] R. Verfürth : A posteriori error estimators for the Stokes equations. Numer. Math. (to appear 1989)

R-Transforming Smoothers
for the Incompressible Navier-Stokes Equations

Gabriel Wittum

Sonderforschungsbereich 123

Universität Heidelberg

Im Neuenheimer Feld 294

D-6900 Heidelberg

Summary

In the present paper we discuss several features of r-transforming smoothers for the incompressible steady-state Navier-Stokes equations in two dimensions. So we compare the influence of different artificial boundary conditions on the numerical performance of the TILU method from [8]. Further, we apply ILU_β from [9] in the r-transforming framework and obtain a reasonable improvement of the SIMPLE method by that. Finally we compare the efficiency of the DGS/TILU method and the SIMPLE methods. All comparisons are based on a standard driven cavity test problem.

1. Introduction

Recently, the use of r-transforming smoothers for the incompressible steady state Navier-Stokes equations has become more and more common in the multi-grid community, as can be seen from [1], [3], [5], [8], [10] and the references there. The introduction of the r-transforming framework in [8] and the theoretical results from [10] already brought certain insight into these smoothers. However, there are still many problems left.

In the present paper we consider several details of constructing a r-transforming smoother for the incompressible, steady-state Navier-Stokes equations

$$-\Delta u + Re \cdot u \cdot \nabla u + \nabla p = f$$
$$\nabla \cdot u = 0 \quad \text{in } \Omega \subset \mathbb{R}^2 \quad (1.1)$$
$$u = g \quad \text{on } \partial\Omega.$$

So we discuss the choice of the artificial boundary condition for the r-transformation. We apply ILU_β from [9] instead of standard ILU, we improve the SIMPLE smoother by this modification and make it more robust and we finally compare the r-transforming smoothers of DGS and SIMPLE type using the standard driven cavity test problem for Reynolds-numbers $0 \leq Re \leq 1000$.

2. R-Transforming Smoothers

We discretize (1.1) on staggered grids as described in [2]. The resulting linearized discrete operator

$$K_\ell(u_\ell^0) = \begin{pmatrix} Q_\ell(u_\ell^0) & \nabla_\ell \\ -\nabla_\ell^T & 0 \end{pmatrix} \tag{2.1a}$$

with

$$Q_\ell(u_\ell^0) = -\Delta_\ell + N_\ell(u_\ell^0), \; N_\ell(u_\ell^0) = u_\ell^0 \cdot \nabla_\ell + \nabla u_\ell^0 \cdot I_\ell \tag{2.1b}$$

is indefinite. Thus a classical splitting with

$$K_\ell(u_\ell^0) = M - N, \; M \text{ regular, "easily invertible"} \tag{2.2}$$

is not possible. A simple remedy is to use a r-transforming iteration for smoothing as introduced in [8]. To that end we construct an auxiliary matrix $\overline{K}_\ell$ such that a splitting

$$K_\ell \overline{K}_\ell = M - N \tag{2.3}$$

makes sense. The corresponding r–transforming iteration to solve

$$K_\ell x = b \tag{2.4}$$

reads

$$x^{(i+1)} = x^{(i)} - \overline{K}_\ell M^{-1}(K_\ell x^{(i)} - b). \tag{2.5}$$

Two r-transformations yield common methods for (1.1). First the DGS and TILU-method as introduced in [1] and [8] using

$$\overline{K}_\ell = \begin{pmatrix} I_\ell & \nabla_\ell \\ 0 & -Q'(u_\ell^0) \end{pmatrix} \tag{2.6}$$

with

$$K_\ell \overline{K}_\ell = \begin{pmatrix} Q_\ell(u_\ell^0) & W \\ -\nabla_\ell^T & -\Delta_\ell^N \end{pmatrix}, \; W = Q_\ell(u_\ell^0)\nabla_\ell - \nabla_\ell Q_\ell'(u_\ell^0) \tag{2.7}$$

second the SIMPLE methods, introduced in [4] using

$$\overline{K}_\ell = \begin{pmatrix} I_\ell & -D_\ell^{-1}\nabla_\ell \\ 0 & I_\ell \end{pmatrix} \tag{2.8}$$

where D_ℓ^{-1} is some approximate inverse for $Q_\ell(u_\ell^0)$ and

$$K_\ell \overline{K}_\ell = \begin{pmatrix} Q_\ell(u_\ell^0) & (I_\ell - Q_\ell(u_\ell^0)D_\ell^{-1})\nabla_\ell \\ -\nabla_\ell^T & \nabla_\ell^T D_\ell^{-1}\nabla_\ell \end{pmatrix}. \tag{2.9}$$

The DGS/TILU methods have been originally designed as multi-grid smoothers whereas SIMPLE methods are widely used as iterative solvers in technical production software. In multi-grid context it was first used by Lonsdale, [3], and even together with some turbulence model by Scheuerer et al. in [5].

A crucial point of the realization of SIMPLE is the choice of D_ℓ^{-1}. As discussed in [8] it is reasonable to use a **perturbed r-transforming smoother**. Here we leave the upper right block of $K_\ell \overline{K}_\ell$ for splitting (2.3) but still use the original $\overline{K}_\ell$ for the transformation in (2.5). Additionally, we are free to prescribe boundary conditions for $\overline{K}_\ell$ and $K_\ell \overline{K}_\ell$ provided both operators still are stable and yield a convergent method. The choice of boundary conditions may have considerable influence on the effectivity of the resulting method as discussed below.

One of the main requirements for a multi-grid method which is to be used for fluid-mechanical problems is **robustness**, meaning that the algebraic solver is insensitive to singular perturbations (see [7], [9]). Robust methods for scalar problems can be constructed via incomplete LU-smoothers (ILU). The r-transforming approach allows the extension of these scalar methods to indefinite systems like the Navier-Stokes equations (in [8]). In [9] we proved the robustness of a modified ILU scheme, ILU_β, for an anisotropic model problem. Now we apply that new ILU variant in r-transforming manner to both r-transformations given above.

3. Numerical Results

In [8] we introduced r-transforming ILU for (1.1) combining a standard incomplete decomposition with transformation (2.6). Now we replace the standard ILU by ILU_β from [9]. Further, we compare the r-transformations (2.6) and (2.8). For (2.6) we investigate about the influence of the boundary condition in $\overline{K}_\ell$ and $K_\ell \overline{K}_\ell$ on the convergence of the perturbed r-transforming iteration. For (2.8) we are interested in the influence of the choice of D_ℓ^{-1}.

We handle the nonlinearity of the problem by a simplified Newton method, i.e. a Newton method where the full Jacobian is simplified. In our case we replace $N_\ell(u_\ell^0)$ from (2.1b) by

$$N_\ell(u_\ell^0) = u_\ell^0 \cdot \nabla_\ell. \qquad (3.1)$$

For our tests we used a driven cavity problem in the unit square as done in [8].

As mentioned above, in perturbed transforming methods we are free to choose boundary conditions for $\overline{K}_\ell$ and $K_\ell \overline{K}_\ell$ separately. In our tests we used the following combinations of boundary conditions

$$-\Delta_\ell := h_\ell^{-2} \begin{bmatrix} & -1 & \\ -1 & a(x_h, y_h) & -1 \\ & -1 & \end{bmatrix} \qquad (3.2a)$$

with

$$a(x_h, y_h) := \begin{cases} 4 & \text{for } h_\varrho < x_h, y_h < 1-h_\varrho \\ 3 & \begin{cases} \text{for } x_h \in \{h_\varrho, 1-h_\varrho\}, h_\varrho < y_h < 1-h_\varrho \\ \text{for } y_h \in \{h_\varrho, 1-h_\varrho\}, h_\varrho < x_h < 1-h_\varrho \end{cases} \\ 2 & \text{else} \end{cases} \quad (3.2b)$$

corresponding to a v. Neumann condition or

$$a(x_h, y_h) := \begin{cases} 4 & \text{for } h_\varrho < x_h, y_h < 1-h_\varrho \\ 5 & \begin{cases} \text{for } x_h \in \{h_\varrho, 1-h_\varrho\}, h_\varrho < y_h < 1-h_\varrho \\ \text{for } y_h \in \{h_\varrho, 1-h_\varrho\}, h_\varrho < x_h < 1-h_\varrho \end{cases} \\ 4 & \text{else} \end{cases} \quad (3.2c)$$

representing a special stencil which is constructed to minimize the entries in the perturbation W from (2.7) for $Re = 0$, or

$$a(x_h, y_h) := \begin{cases} 4 & \text{for } h_\varrho < x_h, y_h < 1-h_\varrho \\ 5 & \begin{cases} \text{for } x_h \in \{h_\varrho, 1-h_\varrho\}, h_\varrho < y_h < 1-h_\varrho \\ \text{for } y_h \in \{h_\varrho, 1-h_\varrho\}, h_\varrho < x_h < 1-h_\varrho \end{cases} \\ e & \text{else} \end{cases} \quad (3.2d)$$

where e is an input parameter for the corresponding routine. The case $e = 6$ corresponds to a Dirichlet boundary condition on the p-grid. Using (3.2b) in $\overline{K}_\varrho$ we obtain for $Re = 0$

$$W = (W_u, W_v)^T, \quad (3.3)$$

$$W_u = \begin{pmatrix} \mathfrak{w}_u & & & 0 \\ & \ddots & & \\ 0 & & & \mathfrak{w}_u \end{pmatrix}, \quad \mathfrak{w}_u := 2 h_\varrho^{-3} \begin{pmatrix} -1 & 1 & & & \\ -1 & & 1 & & \\ & & \ddots & & \\ & & & -1 & 1 \end{pmatrix}$$

and

$$W_v = \begin{pmatrix} -\mathfrak{w}_v & \mathfrak{w}_v & & & \\ -\mathfrak{w}_v & & \mathfrak{w}_v & & \\ & & \ddots & & \\ & & -\mathfrak{w}_v & \mathfrak{w}_v \end{pmatrix}, \quad \mathfrak{w}_v := 2 \cdot h_\varrho^{-3} \begin{pmatrix} 1 & & & \\ & 0 & & \\ & & \ddots & \\ & & & 0 \\ & & & & 1 \end{pmatrix}.$$

With $\overline{K}_\varrho$ from (3.2c) W becomes

$$W_u = h_\varrho^{-3} \text{ blockdiag}\{0, \mathfrak{w}'_u, \ldots, \mathfrak{w}'_u, 0\}, \quad \mathfrak{w}'_u := \begin{pmatrix} 2 & & & \\ & 0 & & \\ & & \ddots & \\ & & & 0 & -2 \end{pmatrix}, \quad (3.4)$$

and

$$W_v := 2 \cdot h_\varrho^{-3} \begin{pmatrix} \mathfrak{w}'_v & & & \\ & 0 & & \\ & & \ddots & \\ & & & 0 & -\mathfrak{w}'_v \end{pmatrix}, \quad \mathfrak{w}'_v := \begin{pmatrix} 0 & & & \\ & 1 & & \\ & & \ddots & \\ & & & 1 \\ & & & & 0 \end{pmatrix}.$$

From fig. 1 it can be seen that the results strongly depend on the input parameter e. There we show the residual convergence factor κ_{10} averaged over ten iterations for the Stokes problem on a finest grid with stepsize $h = 1/256$ versus the modification parameter.

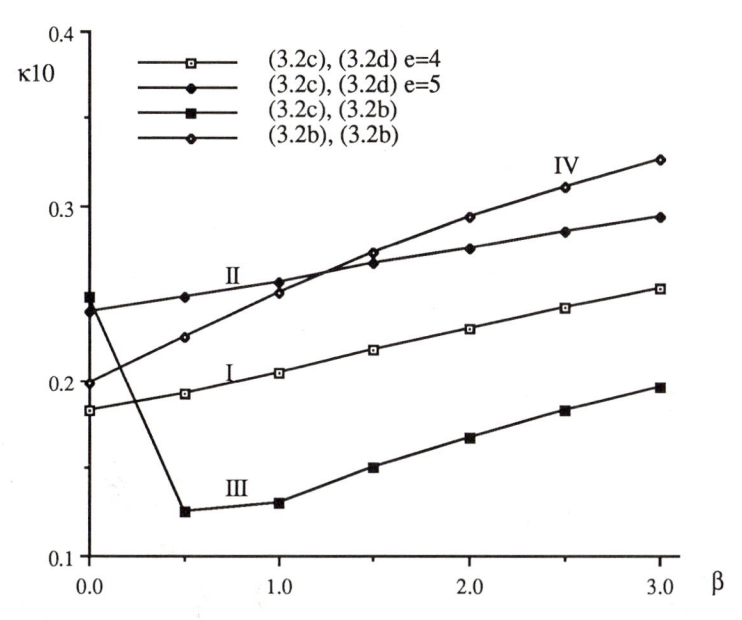

Figure 1: TILU$_{(2.6)}$: The convergence factor κ_{10} versus modification parameter β for $Re = 0$ with varying boundary conditions for $\overline{K}_\varrho$ (first mentioned) and $K_\varrho \overline{K}_\varrho$ (second). The multi-grid parameters are: V-cycle, 2 smoothing steps, starting value by nested iteration with one mg-call per level.

The artificial boundary condition for $\overline{K}_\varrho$ and $K_\varrho \overline{K}_\varrho$ serves as parameter. Four combinations of boundary conditions are shown. The most effective one is the combination (3.2c) for $\overline{K}_\varrho$ and the Neumann stencil (3.2b) for $K_\varrho \overline{K}_\varrho$ (curve III). This branch shows a distinct minimum at $0.5 < \beta_{opt} < 1$. For $\beta > \beta_{opt}$ κ_{10} increases. Curves I and II are obtained using (3.2c) for $\overline{K}_\varrho$ and (3.2d) for $K_\varrho \overline{K}_\varrho$. They differ only by setting $e = 4$ (curve I) or $e = 5$ (curve II) resp. Both can be viewed as continuation of curve III for larger values of β. In order to explain that we have to consider the smoothing property of TILU$_\beta$ for the Stokes equations. In [10] we proved that in the unperturbed case the smoothing property for the whole system is given by the one for the two diagonal blocks. For these Poisson problems we analyzed ILU$_\beta$ in [9] and obtained a dependency of κ on β similar to curve III. In [9] we determined for the optimal β the value ≈ 0.45. Here we have $0.5 < \beta_{opt} < 1$. This is due to the perturbation W as will be discussed in detail in a forthcoming paper. The analysis in [9] shows that the increase of κ for $\beta > \beta_{opt}$ is caused by the growing norm of the rest matrix N from (2.2). Increasing the diagonal entries in $-\Delta_\varrho^N$ in $K_\varrho \overline{K}_\varrho$ from (2.7) has the same effect for fixed β. The corner value e plays a special rôle during the decomposition process since denoting by d_{ij} the elements of the diagonal matrix D of the incomplete decomposition with $M = (L+D) D^{-1} (U+D)$, M from splitting (2.2) we have

$$e = \begin{cases} \max\{d_{ij}\} & \text{for } \beta \in [0,1], e \geq 4 \\ \min\{d_{ij}\} & \text{for } \beta > 1, e \leq 4 \end{cases} \quad (3.5)$$

Thus it has a rather important influence on M^{-1} too. This explains the surprising difference between curve I and curve II.

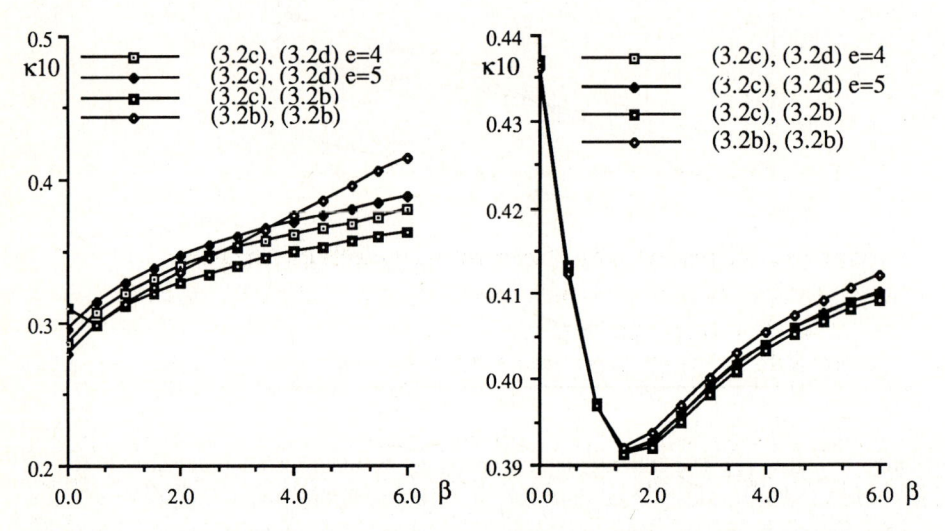

Figure 2: TILU$_{(2.6)}$: Results for $Re=100$

Figure 3: TILU$_{(2.6)}$: Results for $Re=1000$

Figs. 2 and 3 show the same algorithm for larger Re. There we have a stronger influence of W as it increases linearly with Re. We see that larger β serves to compensate this increasing perturbation. Further the difference between the several boundary conditions for the diffusion terms vanishes as the large nonlinear terms dominate.

Next we applied the SIMPLE transformation (2.8). There we used as approximate inverse D_ℓ^{-1} with

$$D_\ell^{-1} := \text{diag}\{\alpha \cdot q_{ii}\} \quad (3.6)$$

with q_{ii} the diagonal element of $Q_\ell(u_\ell^0)$ and $\alpha \in \mathbb{R}$ some parameter.

as originally proposed with $\alpha=1$ by Patankar and Spalding, [4]. This must be combined with damping in iteration (2.5) using different damping parameters $\omega_{u/v}$ and ω_p for velocities and pressure. There are also many variants like the enhanced pressure correction scheme by van Doormaal and Raithby, [6], which looks more complicated but corresponds in the present case to $\alpha > 1$; further the SIMPLER, SIMPLEST, PISO schemes etc. They all

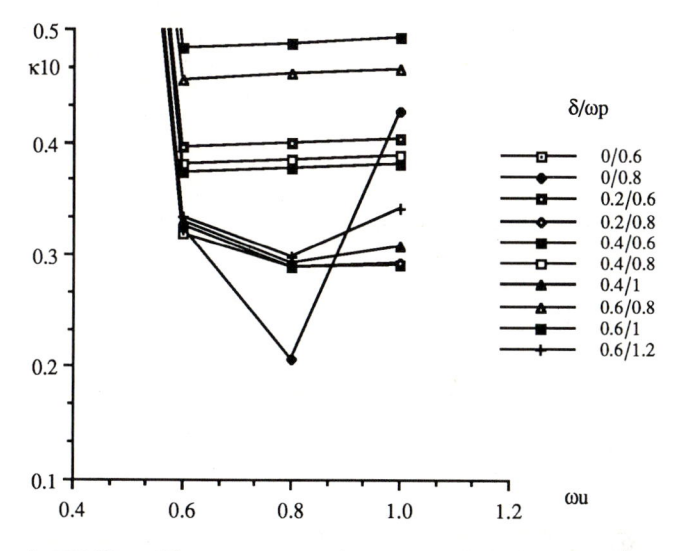

Figure 4: TILU$_{(2.8)}$: The convergence factor κ_{10} versus damping parameter ω_u for $Re = 0$ and several combinations of δ and ω_p. The multi-grid parameters are: V-cycle, 2 smoothing steps, starting value by nested iteration with one mg-call per level. The parameter δ is given by $\delta := \alpha - 1$.

try to improve the approximate inverse D_ℓ^{-1} in (2.8) and (2.9). This is important if we want to have a solver. However, we only are seeking for a smoother and there a main requirement is simplicity. Thus, we restrict to (3.6). The resulting product system (2.9) is again factorized by ILU. Fig. 4 shows the convergence factor κ_{10} versus the damping parameters $\omega_u = \omega_v$ and ω_p for a grid with finest stepsize $h=1/64$. Finer grids gave analogous results.

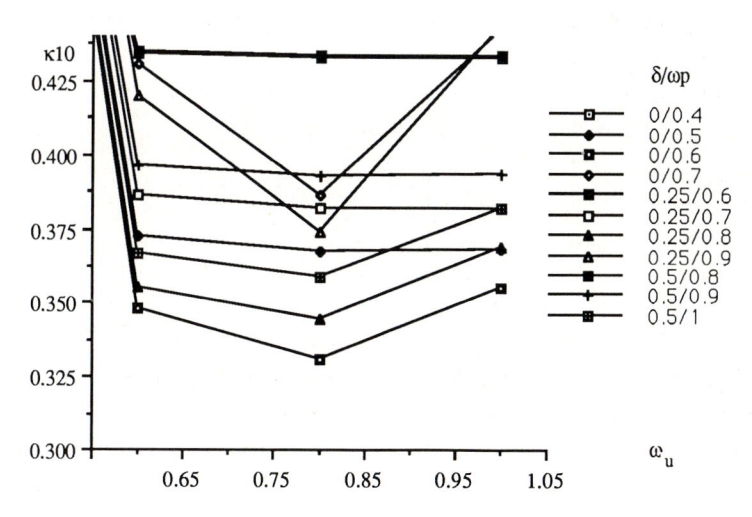

Figure 5: TILU$_{(2.8)}$: Results for $Re = 100$. Finest stepsize: $h = 1/32$, $\delta := \alpha - 1$.

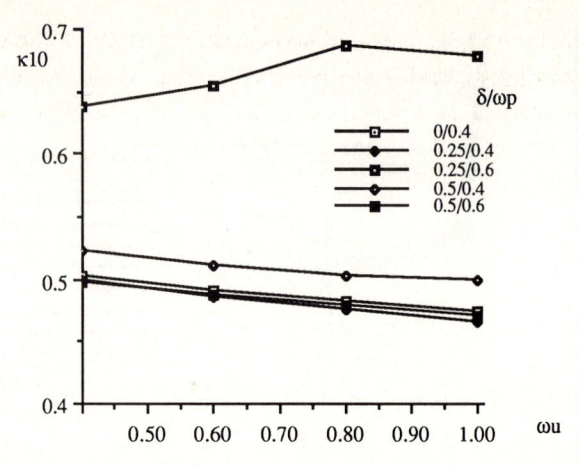

Figure 6: TILU$_{(2.8)}$: Results for $Re = 1000$. Finest stepsize: $h = 1/128$, $\delta := \alpha-1$.

We see that the method can be quite effective, but strongly depends on the right choice of parameters and there are three ones. For higher Re as shown in figs. 5 and 6 the sharpness of this dependency slackens. However, the method deteriorates too. For $Re = 1000$ the convergence factors are comparable to TILU$_{(2.6)}$. The main problem is that we have to adjust a couple of damping parameters appropriately.

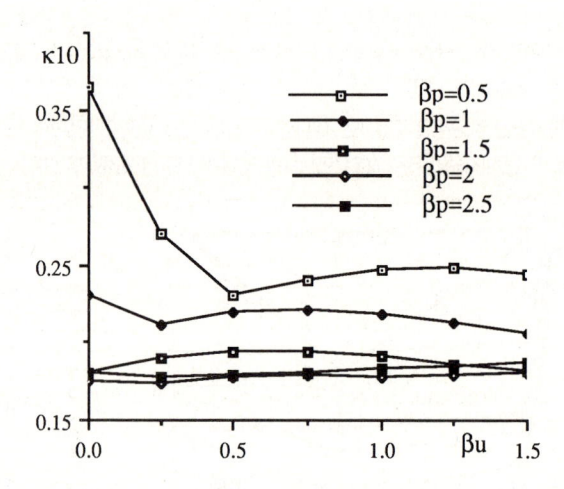

Figure 7: TILU$_{(2.8)}$: The convergence factor κ_{10} versus modification parameter β_u for $Re = 0$ and several values of β_p. The multi-grid parameters are: V-cycle, 2 smoothing steps, starting value by nested iteration with one mg-call per level. The parameter δ is given by $\delta := \alpha-1$.

In [11] we showed that damping is equivalent to modifying an iterative method; i.e. to use ILU_β from [9] instead of standard ILU. Thus we tried to avoid the usual 3 parameters α, ω_u, ω_p which have to be adjusted very carefully and instead use ILU_β with two different modification parameters β_u and β_p while leaving $\alpha = \omega_u = \omega_p = 1$. The corresponding results are given in figures 7–9 for different Re.

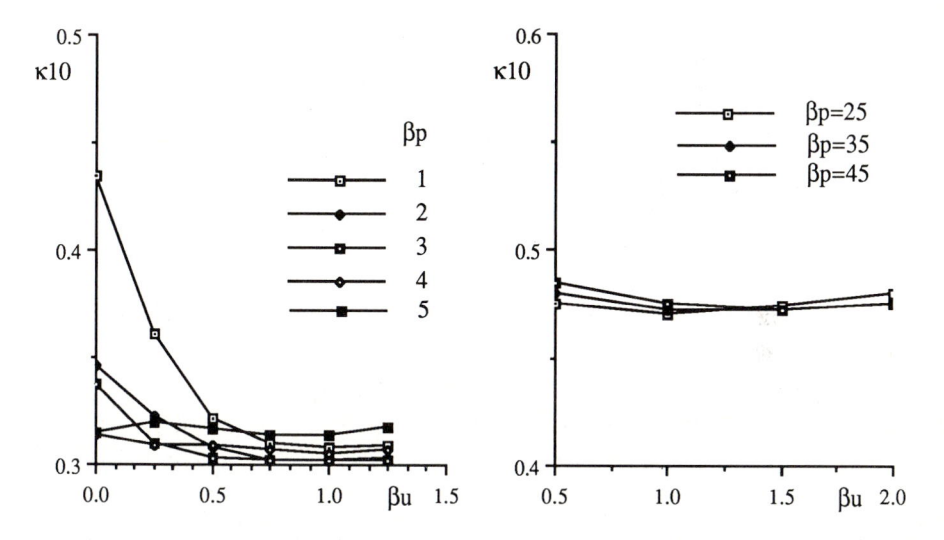

Figure 8: Corresp. results for $Re = 100$.

Figure 9: Corresp. results for $Re = 1000$.

We clearly see that provided $\beta_u \geq 0.5$ and $\beta_p > 1$ the method is very effective and there is no need of exactly adjusting one special pair of parameters. Choosing $\beta_u = 1$ and $\beta_p = \beta_p(Re)$ large enough is sufficiently near to the optimal result. Reasonable values for β_p from figures 7–9 are

$$\beta_p = \begin{cases} 2 & Re = 0 \\ 3 & Re = 100 \\ 25 & Re = 1000 \end{cases} \quad (3.7)$$

This result corresponds to the results of the comparison between damping and modification in [11].

Finally, figure 10 compares the convergence factors for the optimal $TILU_\beta$ strategy with both transformations (2.6) and (2.8). We see that transformation (2.6) gives results which are slightly superior. Further, the amount of work to compute the ILU_β factorization with transformation (2.8) needs additional $7N$ operations to compute D_ϱ and one additional word of memory per grid-point to store D_ϱ while the iteration step of both methods needs the same amount of work. Thus transformation (2.6) yields the more effective method.

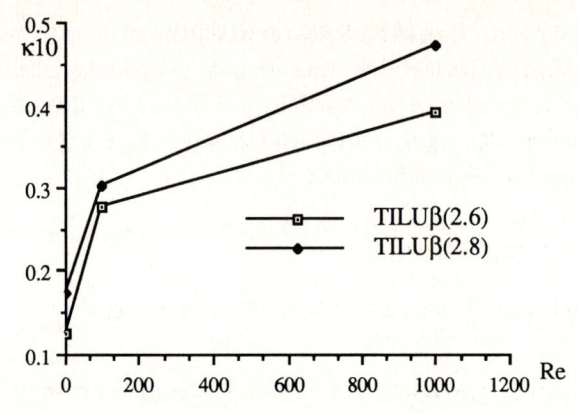

Figure 10: Comparison between the optimal results of $\mathrm{TILU}_{\beta,(2.6)}$ and $\mathrm{TILU}_{\beta,(2.8)}$.

References

[1] Brandt,A, Dinar,N: Multigrid solutions to elliptic flow problems. ICASE Report Nr. 79-15 (1979).

[2] Harlow,F.H., J.E. Welch : Numerical Calculation of time-dependent viscous incompressible flow of fluid with free surface. The Physics of Fluids 8,12 (1965),2182-2189.

[3] Lonsdale,G. : Solution of a rotating Navier-Stokes problem by a nonlinear multigrid algorithm. Report Nr. 105, Manchester University, (1985).

[4] Patankar, S.V. , D.B. Spalding : A calculation procedure for heat and mass transfer in three-dimensional parabolic flows. Int. J. Heat Mass Transfer, 15 (1972), 1787-1806

[5] Periç,M., Rüger,M., Scheuerer,G.: A finite volume multigrid method for calculating turbulent flows. TSF 7 Stanford University, Aug. 1989

[6] Van Doormal,J.P., Raithby,G.D.: Enhancements of the SIMPLE method for predicting incompressible fluid flows. Numer. Heat Transf. 7, 147-163 (1984)

[7] Wesseling,P.: Theoretical and practical aspects of a multigrid method. SISSC. 3 (1982), 387-407.

[8] Wittum,G.: Multi-grid methods for Stokes and Navier-Stokes equations.Transforming smoothers - algorithms and numerical results. Numer. Math., 54 , 543-563 (1989).

[9] Wittum,G. : On the robustness of ILU-smoothing. SISSC, 10, 699-717 (1989)

[10] Wittum,G. : On the convergence of multi-grid methods with transforming smoothers. Preprint #468, SFB 123, Universität Heidelberg, 1988

[11] Wittum,G. : Linear Iterations as Smoothers in Multi-Grid Methods. Impact of Computing in Scinece and Engineering , 1,180-215 (1989)

List of Participants

Albrecht, J., Institut für Mathematik, Technische Universität Clausthal, Erzstraße 1, D-3392 Clausthal-Zellerfeld

Ansorge, R., Institut für Angewandte Mathematik, Universität Hamburg Bundesstr. 55, D-2000 Hamburg 13

Artlich, S., SFB 256, Institut für Angewandte Mathematik, Universität Bonn, Wegeler Straße 6, D-5300 Bonn

Axelsson, O., Faculty of Mathematics and Computer Science, University of Nijmegen, Toernooiveld 5, NL-6525 ED Nijmegen

Bader, G., Institut für Angewandte Mathematik, Universität Heidelberg, Im Neuenheimer Feld 293/294, D-6900 Heidelberg

Baensch, E., SFB 256, Institut für Angewandte Mathematik, Universität Bonn, Wegeler Straße 6, D-5300 Bonn

Bastian, P., Institut für Mathematische Maschinen und Datenverarbeitung III, Universität Erlangen-Nürnberg, Martensstraße 3, D-8520 Erlangen

Berger, H., Mathematisches Institut A, Pfaffenwaldring 57, D-7000 Stuttgart

Blum, H., Institut für Angewandte Mathematik, Universität Heidelberg, Im Neuenheimer Feld 293/294, D-6900 Heidelberg

Braess, D., Institut für Mathematik, Ruhr-Universität, D-4630 Bochum

Boorsboom, M., Delft Hydraulics, P.O. Box 152, NL-8300 AD Emmeloord,

Burmeister, J., Institut für Informatik und Praktische Mathematik, Christian-Albrechts-Universität Kiel, Olshausenstraße 40, D-2300 Kiel

Collatz, L., Institut für Angewandte Mathematik, Universität Hamburg, Bundesstraße 55, D-2000 Hamburg 13

Dijkstra, D., Fac. TW, University of Twente, P. O. Box 217, NL-7500 AE Enschede

Dinkler, D., Institut für Statik, Technische Universität Braunschweig, D-3300 Braunschweig

Dörfer, J., Angewandte Mathematik, Universität Düsseldorf, Universitätsstraße 1, D-4000 Düsseldorf 1

Dreyer, T., Institut für Angewandte Mathematik, Universität Heidelberg Im Neuenheimer Feld 293/294, D-6900 Heidelberg

Franzen, H., Ökosystemforschung, Universität Kiel, Schauenburger Straße 112, D-2300 Kiel 1

Fuchs, L., Department of Gasdynamics, The Royal Institute of Technology, S-100 44 Stockholm and Scientific and Technical Computing Group, ACIS IBM Svenska AB, S-163 92 Stockholm

Geiger, A., Rechenzentrum, Universität Stuttgart, Allmandring 30, D-7000 Stuttgart 80

Hackbusch, W., Institut für Informatik und Praktische Mathematik Christian-Albrechts-Universität Kiel, Olshausenstraße 40, D-2300 Kiel 1

Harig, J., Institut für Angewandte Mathematik, Universität Heidelberg, Im Neuenheimer Feld 293/294, D-6900 Heidelberg

Hebeker, F. K., IBM Wissenschaftszentrum, ISAM, Tiergartenstraße 15, D-6900 Heidelberg

Heilmann, M., IBM WIssenschaftszentrum, ISAM, Tiergartenstraße 15, D-6900 Heidelberg

Heinrich, H., Düsternbrooker Weg, D-2300 Kiel 1

Hinder, R., FB Mathematik, Technische Hochschule Darmstadt, Schloßgartenstraße 7, D-6100 Darmstadt

Hoppe, R. H. W., Konrad-Zuse Zentrum Berlin, Heilbronner Straße 10, D-1000 Berlin 31

Horton, G., Institut für Mathematische Maschinen und Datenverarbeitung III, Universität Erlangen-Nürnberg, Martensstraße 3, D-8520 Erlangen

van Kan, J., Dept. of Mathematics and Informatics, Delft University of Technology, Julianalaan 132, NL-2625 BL Delft

Kaps, P., Institut für Mathematik und Geometrie, Universität Innsbruck, Technikerstraße 13, A-6020 Innsbruck

Katzer, E., Institut für Informatik und Praktische Mathematik, Christian-Albrechts-Universität, Olshausenstr. 40, D-2300 Kiel

Kornhuber, R., Konrad-Zuse Zentrum Berlin, Heilbronner Straße 10, D-1000 Berlin 31

Kolodziej, J. A., Institute of Applied Mechanics, Technical University of Poznan, Piotrowo 3, PL-60-965 Pozan

Kröner, D., Institut für Angewandte Mathematik, Universität Heidelberg, Im Neuenheimer Feld 293/294, 6900 Heidelberg

Krukow, G., Abt. EW-403, BMW-AG-München, Postfach 400240,
D-8000 München

Kuerten, J. G. M., Dept. of Applied Mathematics, University of Twente,
P. O. Box 217, NL-7500 AE Enschede

Kuhfuß, R., Rechenzentrum, Universität Stuttgart, Allmandring 30,
D-7000 Stuttgart

Leinen, P., Universität Dortmund, Postfach 500500, D-4600 Dortmund 50

Liebau, F., Institut für Informatik und Praktische Mathematik,
Christian-Albrechts-Universität Kiel, Olshausenstraße 40, D-2300 Kiel 1

Lieser, J., Institut für Aerodynamik und Meßtechnik, TH-Darmstadt,
Petersenstraße 30, D-6100 Darmstadt

MacKenzie, J. A., Computing Laboratory, Oxford University, 8-11 Keble Road,
Oxford, OX1 3QD, England

Martensen, E., Mathematisches Institut II, Universität Karlsruhe,
D-7500 Karlsruhe

Marzi, J., Institut für Schiffbau der Universität Hamburg, Lämmersieth 90,
D-2000 Hamburg 60

Maubach, J., Faculty of Mathematics and Computer Science, University of
Nijmegen, Toernooiveld 5, NL-6525 ED Nijmegen

Mayers, D.F., Computing Laboratory, Oxford University, 8-11 Keble Road,
Oxford, OX1 3QD, England

Müller, B., DFVLR, Deutsche Forschungs- und Versuchsanstalt für Luft- und
Raumfahrt e.V., Institut für Theoretische Strömungsmechanik, Bunsenstr. 10,
D-3400-Göttingen

Mueller, S., SFB 123, Universtität Heidelberg, Im Neuenheimer Feld 293/294,
D-6900 Heidelberg

Piquet, J., ENSM, Laboratoire d´Hydrodynamique Navale, 1 rue de la Noe,
F-44072 Nantes Cedex

Rannacher, R., Institut für Angewandte Mathematik, Universität Heidelberg,
Im Neuenheimer Feld 293/294, D-6900 Heidelberg

Rieger, E., Hermann-Föttinger-Institut für Thermo- und Fluiddynamik,
Technische Universität Berlin, Straße des 17. Juni 135, D-1000 Berlin 12

Rollett, J. S., Computing Laboratory, Oxford University, 8-11 Keble Road,
Oxford, OX1 3QD, England

Rubart, L., Institut für Strömungslehre und Strömungsmaschinen, Universität der Bundeswehr Hamburg, Holstenhofweg 85, D-2000 Hamburg 70

Schäfer, O., Abteilung ZXE 21, ABB Brown Bovery, D-5400 Baden

Schlechtriem, S., Lehr- und Forschungsgebiet Aachen, RWTH Aachen, Templergraben 64, D-5100 Aachen

Schmatz, M. A., Messerschmitt-Bölkow-Blohm GmbH, Postfach 801160, D-8000 München 80

Schmidt, A., SFB 256, Institut für Angewandte Mathematik, Universität Bonn, Wegeler Straße 6, D-5300 Bonn

Schüller, A., Gesellschaft für Mathematik und Datenverarbeitung, F1/T, Postfach 1240, D-5205 St. Augustin 1

Schütz, H., Hermann-Föttinger-Institut für Thermo- und Fluiddynamik, Technische Universität Berlin, Straße des 17. Juni 135, D-1000 Berlin 12

Shah, T. M., Computing Laboratory, Oxford University, 8-11 Keble Roed, Oxford, OX1 3QD, England

Steckel, B., Gesellschaft für Mathematik und Datenverarbeitung, F1/T, Postfach 1240, D-5205 St. Augustin 1

Thiele, F., Hermann-Föttinger-Institut für Thermo- und Fluiddynamik, Technische Universität Berlin, Straße des 17. Juni 135, D-1000 Berlin 12

Turek, S., SFB 123, Universität Heidelberg, Im Neuenheimer Feld 293/294, D-6900 Heidelberg

Verfürth, R., Institut für Angewandte Mathematik, Universität Zürich, Rämistraße 74, CH-8001 Zürich

Wittum, G., SFB 123, Universität Heidelberg, Im Neuenheimer Feld 293/294, D-6900 Heidelberg

Wolter, D., Hermann-Föttinger-Institut für Thermo- und Fluiddynamik, Technische Universität Berlin, Straße des 17. Juni 135, D-1000 Berlin 12

Addresses of the editors of the series "Notes on Numerical Fluid Mechanics":

Prof. Dr. Ernst Heinrich Hirschel (General Editor)
Herzog-Heinrich-Weg 6
D-8011 Zorneding
Federal Republic of Germany

Prof. Dr. Kozo Fujii
High-Speed Aerodynamics Div.
The ISAS
Yoshinodai 3-1-1, Sagamihara
Kanagawa 229
Japan

Prof. Dr. Bram van Leer
Department of Aerospace Engineering
The University of Michigan
Ann Arbor, MI 48109-2140
USA

Prof. Dr. Keith William Morton
Oxford University Computing Laboratory
Numerical Analysis Group
8-11 Keble Road
Oxford OX1 3QD
Great Britain

Prof. Dr. Maurizio Pandolfi
Dipartimento di Ingegneria Aeronautica e Spaziale
Politecnico di Torino
Corso Duca Degli Abruzzi, 24
I-10129 Torino
Italy

Prof. Dr. Arthur Rizzi
FFA Stockholm
Box 11021
S-16111 Bromma 11
Sweden

Dr. Bernard Roux
Institut de Mécanique des Fluides
Laboratoire Associè au C.R.N.S. LA 03
1, Rue Honnorat
F-13003 Marseille
France